PERSPECTIVES ON THE ENVIRONMENT

Perspectives on the Environment

Interdisciplinary research in action

Interdisciplinary Research Network on Environment and Society

Edited by
JANE HOLDER
PAULINE LANE
SALLY EDEN
RACHEL REEVE
UTE COLLIER
KEVIN ANDERSON

Avebury
Aldershot · Brookfield USA · Hong Kong · Singapore · Sydney

Published by Avebury
Ashgate Publishing Limited
Gower House
Croft Road
Aldershot
Hants GU11 3HR
England

Ashgate Publishing Company
Old Post Road
Brookfield
Vermont 05036
USA

A CIP catalogue record for this book is available from
the British Library

Typeset by Jeff Munn-Giddings
60 Chapel Road, West Bergholt
Essex CO6 3HZ

ISBN 1 85628 606 1

Printed and Bound in Great Britain by
Athenaeum Press Ltd, Newcastle upon Tyne.

Contents

Part III European Environmental Policy: A Comparative Assessment

Part IV Urban Environmental Problems: East and West

Acknowledgements

Whilst recognising the support and advice of many people, we would particularly like to thank all the contributors for their papers, Andy Dobson for writing the foreword, Jeff Munn-Giddings for typesetting, Jevon White for computer assistance, Ted Benton for his guidance and encouragement, the ESRC for their financial support of IRNES from 1991 - 1993 and last, but not least, the continued interest in interdisciplinary research expressed by the growing and active membership of IRNES.

Declaration

The views expressed within the individual papers of this book are those of the respective authors and do not necessarily reflect either the views of IRNES or the ESRC.

Acknowledgements

[illegible]

Disclaimer

[illegible]

Foreword

Professor Andrew Dobson

It is hard to imagine a field of study more suited to interdisciplinary research than the environment. The standard intellectual disciplines deal with particular facets and features of the human condition as though they were separate and discrete, and the tide of intellectual endeavour since the Renaissance has flowed steadily in this direction. Nothing that has happened internal to these disciplines has done much to bring the natural sciences, social sciences and humanities together, and contemporary academic conferences are mostly hermetic and highly specialised. In recent years, though, it has become ever more clear that human activity takes place within a generalised context that sustains and constrains it, and that the nature of this sustaining and constraining is a subject of crucial and enduring interest.

It is often the case that something comes to be of interest when we can no longer take it for granted, and the environment is no exception. Concern for the environment is of course nothing new: public health, for example, has been associated with environmental conditions for a very long time, and as such the environment has been an object of public policy for over 600 years – in Britain at least. Similarly, the wanton exploitation of North America's forests in the nineteenth century raised the issue of the environment as a resource, and gave rise eventually to the associated (and now highly contested) idea of sustainability. In general, though, in these times there was hardly a sense that humanity's interactions with the environment were of such a character that they threatened to undermine the possibility of interaction itself. Apart from some select biological and earth scientists in France, Britain, the USA, Russia and Germany whose studies made them aware of the possibility of serious resource and energy scarcity, the general view was that environmental problems were isolated and discrete and could be dealt with without reference either to each other or to the productive nexus within which they were embedded.

What is new about the present environmental problematic is its global nature and reach – at least from a Northern and Western point of view (from the perspective of someone in Ecuador dying for want of clean water, environmental problems look more local and are susceptible, in principle, to local treatment). In this sense the archetypal contemporary environmental problem is global warming, the implications of which are not confined to this or that point in time, but to the whole planet, and for the foreseeable future. This trend towards globalisation and towards extension in general, is matched by the way in which the intellectual disciplines have reacted to it. It is as though the spatial extension of environmental problems has been ghosted by a willingness (born of necessity) on the part of intellectuals to break out of their self-imposed restraints and to talk to people in other disciplines.

Global warming is indeed a case in point. It is remarkable how important science has been to raising environmental awareness over the past 30 years. Like it or not, we work fundamentally within the Enlightenment idiom, and this means that public acceptance of the existence of a problem (susceptible to natural scientific investigation), and mobilisation around it, depends largely upon the blessing of scientists. Rightly or wrongly, scientists believe that they can describe with some precision the physical dynamics that contribute to global warming, and it has been incumbent upon both social scientists and intellectuals working within the humanities to listen to what scientists have to say about global warming. Nothing that social scientists say about the phenomenon will mean much without a clear understanding of its natural scientific base.

At the same time, social scientists (or some of them at least) are by training acutely aware that knowledge is not so much acquired as produced, and that the meaning of global warming (and of 'the environment' in general) is fiercely contested. Scientists are simply one 'epistemic community' among others, from this point of view, and so the way that environmental problems are constructed, and the way in which different publics produce them, are properly the domain of interpretative social science. It is clear from this that natural and social scientists will not always see eye to eye, but the engagement with a common problem that drives their agreements and disagreements is the source of real and fecund forms of interdisciplinarity. Other social scientists, less concerned with interpretation and more comfortable with description, have sought to explain how political systems and the policy-making process have reacted to the particular problems posed by environmental difficulties. In some cases this has merely involved incorporating new data into old explanatory frameworks, while in others the frameworks themselves have come to sound somewhat creaky.

Within the humanities, widely understood, the question raised by the environmental problematic – how should we live? – is not new, but the answers to it must now take account of factors barely considered in the majority of modern political ideologies. Environmental philosophers wonder what our proper relationship with non-human nature should be, and political theorists debate the nature of the just society under conditions of apparently enduring and deepening scarcity. It is remarkable to see how the 'ought' questions that exercise political theorists and philosophers are so informed (in the environmental context) by what 'is', and the information as to what 'is' is derived quite confessedly from disciplines outside the humanities themselves.

The environment as a common object of concern weaves its thread, therefore, through the encounters between intellectual disciplines andtheir practitioners. It seems obvious that solutions to our environmental problems will be found within an interdisciplinary framework, or not at all. But what seems obvious in theory is not always made possible in practice, and organisational energies are a scarcer commodity than wide-eyed plans. The Interdisciplinary Research Network on Environment and Society (IRNES) owes its existence, principally, to the drive and perseverance of the editors of this book. They have seen that we are faced with an interdisciplinary challenge and they have combined to give it a voice. It was a privilege to be asked to adress the first IRNES conference in 1992, and it is a privilege to be invited to write this foreword to this excellent collection of the conference's papers. Here is evidence of interdisciplinarity at work, and I, for one, am looking forward to more.

Professor Andrew Dobson
Keele University
1993

Introduction

There are many new recruits to the environmental research ranks every year. Each brings their own perspective and their own commitment to change academic and practical debates. In the face of this diversity of perspectives, the disciplinary approaches which have dictated philosophies and methodologies of study for many years are recombining. In a period of tightening research budgets, it can often be hazardous to be interdisciplinary, but in the environmental field it is also becoming more rewarding.

This book presents a collection of papers from the recent IRNES 1992 conference as part of an ongoing commitment to interdisciplinary environmental research and to representing the range of perspectives it encompasses. This Introduction will describe the principles on which IRNES is founded, and on which we continue to thrive, and then outline the themes of this book.

IRNES is the acronym for the Interdisciplinary Research Network on the Environment and Society. The network was founded in 1990 expressly to facilitate interdisciplinary liaison between active researchers in the environmental field. From the initial efforts of only a few committed postgraduates, IRNES has developed into a national network of academic, postgraduate and postdoctoral researchers, as well as non-academic workers. We have members from a diversity of backgrounds, including economics, law, geography, sociology, environmental science, agriculture, politics, policy studies and philosophy. This diversity makes IRNES a particularly active and dynamic research network. IRNES is also fully democratic: there are no elected committees or organisers. Instead, members are encouraged to contribute time and effort as they are able, thereby involving as many people as practicable in the planning and administration of IRNES and sharing the burdens of work and responsibility.

Briefly, the main objectives of IRNES are: to facilitate contact between environmental researchers; to encourage postgraduate involvement in interdisciplinary research; to encourage participation from all disciplines; to maintain national coverage; to encourage and disseminate members' individual and collaborative research. Progress towards all five of these aims is maintained through our conferences. The first IRNES conference was held in Leeds on 14th and 15th September 1992, entitled Perspectives on the Environment: Research and Action for the 1990s. This book brings together the best papers from this conference and helps IRNES to demonstrate its continued commitment to promoting and disseminating interdisciplinary environmental research.

We believe it is necessary to broaden the often insular approaches to environmental issues and to develop more robust and comprehensive research strategies to address environmental problems. The interdisciplinarity and range of papers at the 1992 conference was praised by delegates and we feel that this addressed a need in the academic community for such exchange and support across traditional disciplinary boundaries. We received many positive comments, both during and after the conference, which underlined its effectiveness as a national interdisciplinary forum for environmental research.

We will continue to expand the network because, as environmental research continues to expand, it will have an increasingly significant role to play in encouraging and supporting research in all manner of environmental fields.

This book attempts to bring interdisciplinary work stemming from IRNES' activities to a wider audience. The book is divided into four parts, each of which is centred upon a substantive environmental theme rather than a disciplinary or theoretical perspective. The first three parts take the main themes of ideology, regulation and policy as the key areas of active environmental research and the most promising for collaboration across the social sciences. The fourth part offers two contrasting case studies, illustrating the international character of air pollution problems.

Overall, we have sought to encourage new research on key issues which is strongly academic but also shows critical analysis of environmental initiatives and progress so far. This highlights the contribution of environmental research to the practical and political debates about environmental protection in the current national and international agenda.

The first part of the book takes green ideology and its relevance to the political domain as its theme and includes a survey of a key political green group, the British Green Party, and critiques of less prominent ideologies.

Ideologies both underlie our attitudes towards the environment and the non-human world and prescribe out social and political activities for its protection. While 'green' ideology is frequently identified as a single discourse for study, this section demonstrates that the label encompasses a diversity of ideas, which require a critical analysis unfettered by approaches embedded in only one discipline.

Ian Coates, in Chapter 1, considers how the National Front has harnessed particular aspects of green ideology both to widen the appeal of the fascist agenda and to distract attention from their more extremist policies: the NF thus attempted a partial reconstruction of its ideology. Nevertheless, Coates argues that green ideology remained subordinate to the NF's dominant ideological outlook, making a total 'greening' impossible. Rüdig, Franklin and Bennie, in Chapter 2, put into context the fluctuating fortunes of the British Green Party and identify how external factors contributed to its early successes, especially the poor state of the competing parties in the 1989 European election. Conversely, the authors consider the more recent decline of the party's membership, exploring the tensions over internal organisation and how external developments reduced the Green Party's effectiveness in the face of reinvigorated competition.

John Barry, in Chapter 3, delves into the ideological and philosophical foundations of a very different, more loosely delineated, group when he attempts a critical analysis of the contribution of deep ecology to green political theory. Such a contribution produces tensions and Barry points to deep ecology's failure to address issues of social justice that are important for more pragmatical green groups. He argues for an extension of moral value to members of the non-human world, whilst exposing deep ecology's lack of substantial political promise in a practical sense.

From most environmental ideologies stems a need for the protection of the environment from the activities of industry and other bodies. We can therefore move from ideological analyses to more pragmatical considerations of the possible mechanisms for environmental protection. Part II on environmental regulation looks at different ways of regulating developments and industrial activity, with three different sources of pressure. Firstly, legislation is taken as a control strategy, putting industry under pressure from regulatory agencies. Secondly, internal company responsibility is examined as a self-regulation strategy, with pressure from consumers but not from organised bodies. Thirdly, external financial evaluation is proposed as a market strategy, with pressure from investors.

The legislative example of regulation chosen by John Alder, in Chapter 4, is the European Community Directive on Environmental Impact Assessment, which was intended to enable decision makers to make an informed choice between environmental and other objectives and for the public to be consulted in this choice. The UK legal system has failed to emphasise the environmental protection and environmental rights required by the Directive and the regulatory measures are consequently weakly implemented. Looking at a number of important cases, Alder argues that the courts have respected the discretion exercised by administrative authorities and have failed to place sufficient weight on environmental values as against economic interests.

In Chapter 5, Sussi Handberg, Per Christensen and Susse Georg widen the perspective of environmental regulation by considering how clean technology development might necessitate the integration of environmental with health and safety regulatory frameworks and authorities. Their case study brings out the tensions between workers and management in Denmark which may create obstacles to clean technology because of the top-down nature of its introduction. They argue for clean technology to be viewed not merely as a technical process of external regulation but as an issue of internal company changes.

If law, which has long regulated the use of land and resources, is broadly incapable of preventing environmental problems, are non-legal mechanisms of environmental protection more effective? One such non-legal and self-regulatory mechanism is company adherence to a code of environmental responsibility. This would imply the incorporation of environmental and social ethics into corporate decision-making. From field studies of retailing companies, in Chapter 6, Sally Eden asks how far corporate environmental responsibility is premised upon, and driven by, profitability and how far it has external rationales (consumer pressure) as well as internal ones. Her work contributes to the wider debate around the vaunted compatibility of environmental and economic considerations.

An alternative, but as yet untested, approach to regulation is an environmental credit-rating agency, as proposed by Joseph Tanega in Chapter 7. Investment fund managers and traders, as financial players on the capital market, presently disregard environmental performance or only assess it unconsciously. Publication of environmental ratings by an agency would give value to more socially and environmentally responsible organisations and thereby enhance their growth. Tanega sees such an agency as very influential in the global system of capital flows, because of its provision of information direct to financial players. Within a pro-market orientation, he argues that an agency would be more cost-effective, responsive and powerful then regulation because it provides a market incentive for companies to improve their environmental performance and be rated more highly.

Part III looks to the European context of environmental policy to compare developments within different national situations. We begin with two comparative papers and move to a more specific analysis of the EC's impact on one country

Matthew Gandy, in Chapter 8, contrasts British and German experiences of policy in the energy, waste and water sectors. He highlights the importance of emerging political demands for higher environmental standards and therefore a more stringent regulatory framework, which are ranged against mounting concerns about the growing cost of environmental protection. Gandy draws out the role of the state in the two countries and the different ownership patterns of utilities. Demunicipalisation in Germany has heightened the importance of local and regional operations, whereas the intensifying privatisation in the UK has correspondingly demoted the role of local government.

Ute Collier, in Chapter 9, also provides a comparative analysis but concentrates upon the energy sector and global warming. She argues that environmental objectives have not featured high on the agenda, whereas concerns about the security of energy supply and the maintenance of competition have tended to receive more attention. After assessing technical and policy initiatives in the Netherlands, Germany and the UK, Collier highlights some common themes. End-use efficiency remains a cornerstone of global warming strategies in the three countries, resulting in the orientation of current policy towards supply side management. The development of other generating technologies depends on favourable economic evaluations and the extent of monopolies. There is an emphasis upon liberalisation and greater reliance on the market to achieve optimum levels of operation: this weakens the support for a stricter regulatory framework for carbon dioxide emissions.

Moving to the more specific example of Belgium, Marc Hooghe, in Chapter 10, demonstrates that much of the country's environmental policy is a direct response to EC initiatives, despite Belgium's poor record of compliance. He then adopts a modified version of the political opportunity structure approach to remedy the one-sided view that institutional actors (such as the EC) find it easy to influence political decision-making and non-institutional actors (such as pressure groups) don't. Hooghe argues that the use of a common analytical framework allows a clearer assessment of the different obstacles faced by each group of actors in the specific power configuration of Belgian policy.

The final part of the book looks at the substantive issue of urban air pollution in two different contexts: the USA and the former USSR. The two case studies illustrate the diversity and ubiquity of environmental problems as industrial production under different economic and political conditions produces similar problems for health and environmental quality.

The focus for Judith Chernajovsky, in Chapter 11, is Houston, Texas, which over the past century has grown faster than any other US city and relies particularly on the petrochemical industry for its prosperity. Although economic growth is generally accepted as having a beneficial impact on health because of decreasing infant mortality, in Houston, policies of rapid capital accumulation and urban growth have clearly taken place at the expense of both the environment and human health. The author's survey reveals that child illness is more prevalent in one particularly polluted part of the city and cannot be satisfactorily explained by commonly used factors such as socio-economic household attributes.

Urban air pollution is also a serious problem in the former Soviet Union, again due to industrial production. Maria Shahgedanova and Timothy Burt, in Chapter 12, explore the failures of Soviet air pollution regulation, where total emissions are less than in the USA but emissions per unit of GNP are much higher. Here the type of growth is important and distinct from the previous chapter: production quotas under a centralised economy place pressure on industry to produce at any cost. The diverse and fragmented nature of environmental regulatory agencies, the vague legislation and the low political priority of environmental concerns all contribute to poor implementation and enforcement of the multitude of pollution legislation. In the face of the USSR's disintegration, economic collapse and ethnic and military conflicts have further relegated environmental concerns on the list of political priorities.

We can see from the above outline that this first book of IRNES' proceedings is geared to the dissemination of interdisciplinary research. We acknowledge that such research has inherent problems within the current paradigms of disciplinary thinking and critique, as Andrew Dobson has made clear in his Foreword.

As a network, IRNES tries to cushion the fall between disciplines by showing how interdisciplinary approaches can offer insights and by supporting those undertaking such approaches through an informal and congenial setting. With the increasing fragility of funding for various types of research, we wish to strengthen the foundations of interdisciplinary research so that it is not disadvantaged further in the academic competition for recognition. This book marks the consolidation of our first efforts to encourage active, interdisciplinary environmental research. We hope that the papers illustrate the diversity and strengths of this kind of study and that we can urge other researchers to appreciate what interdisciplinary and environmentally conscious research can offer the academic community.

Part I
GREEN IDEOLOGY

1 A cuckoo in the nest: The National Front and green ideology

Ian Coates

The growth in popular environmental awareness occurring over the last 20 years has encouraged a variety of groups to try to establish their green credentials. In the process of being adapted to existing ideologies and interests, environmental concern has taken a variety of forms, not all of which could necessarily be described as green. The ideology of the Green movement remains one measure with which other environmental outlooks can be compared. Yet there are groups who oppose much of what the Greens stand for, but still seek to portray themselves as green. Perhaps the most extreme manifestation of this phenomenon has been provided by a number of nationalist, racist and neo-Nazi groups throughout Europe who claim to have adopted ecological concerns as part of their political outlook. The 'greening' of the National Front in Britain between 1980 and 1990 offers a particular example through which to investigate the extent to which green ideas can be appropriated by groups with a perspective contrary to that of the Green movement. The NF's underlying motivations and the process by which they adopted environmental concerns will be examined in terms of their understanding of green issues and how these were articulated in combination with other elements of their ideology. A comparison with the ideology of the Green Party will be used to investigate the relationship between conceptions of nature and social location, highlighting the extent to which the worldviews of these two groups diverge.

Prior to the 1980s the NF had shown little interest in the environment, issues of race and nation remaining their main preoccupations. However after some initial successes in the early 1970s the NF were facing a period of decline having proved unable to sustain membership levels or electoral support. This decline in popularity can be interpreted in terms of Billig's analysis of the NF's core and surface ideologies [1]. Whilst their surface ideology of opposition to black immigration was aimed at generating mass support, the core ideology of the NF's leadership was based on racial nationalism, biological determinism, anti-semitism and a conspiratorial worldview, reflecting a covert admiration for Nazi Germany [2]. Such a stance was hardly likely to attract many potential supporters, and this, combined with the risk of prosecution for inciting racial hatred, constrained the NF from openly expressing their full beliefs. This created tensions within the NF over how to present their policies for public consumption, leading to frequent leadership struggles and splits.

The NF also faced increasing levels of public opposition to their racist policies, as well as problems caused by the rightward shift of the Conservative Party after 1975 [3]. By taking a tougher stance on immigration, the Conservatives undermined the NF. Faced with a decline in support, and following their poor performance in the 1979 General Election, a group of young activists began moves to take over the NF's leadership, aiming to diversify into new strategic and policy directions [4]. In this they imitated developments amongst European extreme right groups who were attempting to construct a new style of fascism. The NF's young activists were particularly influenced by the Italian group Terza Posizione (Third Position), who claimed to stand, 'between capitalism and communism'. In an attempt to appeal to youth culture Terza Posizione portrayed themselves as national revolutionaries, opposed to the state, imperialism, materialism, consumerism, and urbanism [5]. The incorporation of environmental concerns was a central component of this approach.

The 'greening' of the National Front

Following the Italian government's clampdown on the extreme right after the 1980 Bologna bombing, a number of Terza Posizione members fled to Britain where they established close links with the future leadership of the NF [6]. Using their editorial control of the NF's publications [7], this group of activists began to outline a new ideological stance, including a strong focus on environmental issues. A racial emphasis was immediately apparent in articles on this topic, beginning with the contention that threatened family farms should be safeguarded on the grounds that, 'No section of Britain's community is racially more healthy' [8]. The headline, 'Alien Bankers Destroying British Countryside' [9], blamed 'Zionist' bankers for financing agribusiness and causing environmental destruction. The NF proposed to expropriate farms owned by big business and redistribute them to the British people who would be encouraged to use organic and sustainable farming techniques [10]. An article entitled, 'Blood and Soil', declared, 'Let Britain again be a nation of crofters and free yeoman farmers' [11].

Another element in the NF's argument was that whilst, 'racial pollution is the worst contamination of all' [12], there was also a spiritual pollution, caused by materialism and consumerism [13]. The passive conformity of the urban population was blamed for perpetuating the capitalist system. A mass movement back to the land to live in harmony with nature would, it was claimed, 'restore healthy instincts', and thus, 'defend the community and land from the alien' [14]. The 'repatriation' of the non-white population was also seen as allowing the transformation of depopulated urban areas into villages surrounded by farms and woodland[15]. Technology was not be rejected though: automated industry situated in rural areas was to provide for the population relocated from urban areas[16]. Meanwhile it was proposed that, 'the NF should aim to establish a rural community comprised solely of Nationalists and their families who wish to live a racially aware earth-rooted life style', to provide a concrete example of what a future society would be like under an NF government[17].

Whilst the presence of ecological themes in the context of discussions of lifestyle and technology reveals apparent similarities with other visions of an ideal future rural society[18], the NF's utopia differed from these in that it was based on racial assumptions. What emerged was a backward looking rural nationalism that saw a return to the land as the answer to a decadent modern urban society. Through contact with the soil the British people were to recover their spiritual and racial identity. These articles were an indication of future NF strategy and policy. Attempts were made to redistribute members to rural areas, whilst new technology was used to provide communications and income[19]. Rural locations also provided secure and private surroundings to train new members.

By 1984, with the group advocating this approach in complete control of the NF[20], environmental issues were given even greater prominence. An editorial in their main journal, Nationalism Today, entitled, 'On the Green Front', declared that to the three main premises of the NF's ideology - social justice, racial purity and national soveriegnty - must now be added a fourth; ecology. It continued:

> This war which man is now waging against nature is not only foolhardy, it is downright dangerous. We can't go on polluting the atmosphere, we can't go on poisoning the sea, we can't go on slaughtering our wildlife forever. Sooner or later, future generations will have to pay for the folly of ignoring nature. Are we going to bequeath to future generations of our Race a world which is drained of wildlife and poisoned with chemicals? Man lives in far closer unity with nature than he realises. Our mother earth has been ruthlessly exploited and poisoned for far too long, and no organism can live forever without nourishment. It is up to us to build a Britain in which our people will once again live in harmony with nature. It is up to us to destroy the capitalist System which is not only enslaving the people but is also destroying the land they live on. Finally, it is up to us to build up a Britain where the people are united by the eternal bonds of blood and soil[21].

This could almost have been mistaken for a radical green polemic but for the patriachal language and references to 'Britain', 'Race', and 'blood and soil'. 'On the Green Front' became a regular feature, containing short articles on environmental issues, mostly on topics like acid rain, nuclear waste, and the destruction of hedgerows and woodland. These tended to be rather shallow and repetitive, failing to examine such issues in depth or in a wider context[22].

After yet another leadership split in mid 1986[23], some new themes became apparent. 'Death on the Amazon'[24] discussed threats to both the rainforest and the tribal peoples living there. Though the NF had mentioned the greenhouse effect in a European context before, this was the first time they had discussed the environment in global terms, referring not only to a region beyond the European sphere but also indicating support for an indiginous and non-white population. This reflected further shifts in policy in which the NF declared itself to be separatist but non-racist[25], supporting the rights of all peoples and cultures. However they still favoured 'repatriation' for all non-whites and did not advocate any kind of global response to environmental problems. This championing of diversity intended to leave the peoples of the South to fend for themselves, regardless of the legacy of colonialism and economic exploitation[26].

With the high profile given to the environment following Prime Minister Margaret Thatcher's apparent conversion to the cause in 1988[27], green issues even reached the pages of National Front News, a newspaper aimed at ordinary members. After years of articles almost exclusively on opposition to black immigration, members were suddenly asked to consider the plight of threatened butterflies and dolphins. The final issue even had a front cover story on the Amazon rainforest[28].

Motives and strategies

One way to examine the NF's interest in green issues is in terms of their underlying motives. In the context of earlier references to the problems facing the NF, the articulation of green concerns offered them a chance to diversify into a new policy area seen as popular with wide sections of the public. To an extent it also enabled the NF to tone down their extreme image and distract attention from their underlying policies, as part of a deliberate reconstruction of their ideology. By this means they sought to distance themselves from the historical realities of fascism, claiming to represent an alternative historical tradition that could be portrayed as racially nationalist and green. At first lesser known figures such as Julius Evola, Corneliou Codrianou, Gregor and Otto Strasser and Walther Darrß were resurrected for this purpose in a way that suggested that they represented an oppositional current within fascism[29]. Later, the emphasis was shifted away from fascist sources to show the NF as heirs of a British rural tradition by claiming writers such as William Cobbett, William Morris, Hillaire Belloc and GK Chesterton as major influences[30]. The NF interpreted their work to illustrate the benefits of an idealized Medieval society. In addition the NF evoked notions of England's mythical and pagan past, based on sources that ranged from Arthurian romance to the fantasy novels of JRR Tolkien[31], a change of emphasis which took symbolic form when a stylized Celtic cross replaced the Union Flag as the NF's emblem[32].

Another motivation for going green was provided by the publicity it generated, with the press finding a 'green' NF good copy [33]. Though opportunism must remain a prime factor in motivating the NF to make these changes there were signs that they began to believe their own rhetoric. If the letters pages of the NF's publications were anything to go by green issues proved popular with the membership. It gave them something positive to campaign on. In the words of one article it showed that they were:

> not just an anti-immigration pressure group but a Movement with a complete worldview. We have shown that we are committed to animal welfare, to the protection of our environment ..... However, it is important to emphasise both to the public and to NF members that our campaigns on these topics are not just opportunistic; it is important to show the depth of our commitment [34].

Activists were encouraged to join campaigning groups relating to such issues as ecology or animal rights in order to assert the NF's ideological position on these matters. NF members joining such groups were also to prospect for likely converts and collect intelligence on opponents [35].

The NF also began to note parallels existing between their own viewpoint and that of other groups. As one statement put it:

> This development, where even people who are hostile to us accept many Nationalist ideas, can also be seen in the Green movement, whose anti-materialism and concern for conservation and the quality of life lead ultimately to our ideological position. If individual species of whale are worth preserving, so too are the different human races. If primitive tribes have a right to preserve their culture, so too do the nations of Europe [36].

However, on this last point the NF remained critical of the Greens, precisely because the latter did not see ecology in racial terms.

It was admitted that, 'over the years the NF have developed a ghetto mentality', and failed to note that 'genuine if misguided opposition exists outside the NF'. Having focused on groups like Greenpeace, the Green Party, Friends of the Earth and the various animal welfare organisations the NF argued that whilst:

> not .... all of these groups are 100% sound in terms of members and objectives .... such groups possess both decent people and useful ideas .... our strategy ought to be geared to correcting their illogical, unfounded or anti-British views - to show them that their true home is within the Revolutionary Nationalist movemement [37].

So rather than seeing Greens as the enemy [38], they were now to be regarded as potential allies with whom it was necessary to open up a dialogue, if only to convert them to the NF's position.

After the Green Party's success in the 1989 European Elections [39], the NF highlighted the challenge this constituted to the left/right political spectrum. This was something the NF had also sought to do by representing themselves as, 'Third Positionists', opposed to capitalism and communism. Green politics were thus recognized as offering the NF an opportunity to popularize its new policies, such as:

> Rural distributism, political and economic decentralization, the break up of our cities, the world seen as a patchwork of diverse peoples; all point to a coherent vision of a new society with undoubted appeal to many of those attracted to Green politics ..... it will allow us to show that only we represent their position in its totality ..... the wide range of issues raised by the environmental crisis bring before us many organizations worthy of the assistance of our activists. Nationalists everywhere should ensure that they are in the vanguard of these groups, even to the extent of establishing them where none presently exist[40].

Such an attempt to set up a popular campaigning green organisation with no apparent connection with the NF was already under way, following the establishment of Greenwave in November 1988[41].

Having adopted the name of a legitimate pressure group[42], Greenwave attempted to form links with other environmental organizations and to attract broader support. However Greenwave's links with the NF soon became public knowledge. National Front News carried adverts for Greenwave, NF members were seen distributing its leaflets, and a supporter mentioned its magazine Green Dawn, was an NF official[43]. In April 1989 an NF activist stood as a Greenwave candidate in the Havering by-election[44]. It was also reported that Greenwave's co-ordinator was the hostess at a party held to celebrate the centenary of Hitler's birth[45]. All this somewhat contradicted Greenwave's denials that they were an NF front. Greenwave's efforts were to come to nothing for by this time the NF were in serious trouble. From a membership of 10,000 in 1979 the NF had fallen to 4,000 in 1980 and 3,000 in 1984. After the 1986 split they were down to 1,500 and by January 1989 they numbered only 6-800[46]. By the end of 1989 the NF had totally collapsed, splitting into two small factions; Third Way and the International Third Position/Liberation[47].

The NF's change of strategy had failed for a number of reasons. In 1984 the electoral path had been abandoned and a purge of the old membership instituted. The intention had been to create a core of highly committed activists by targeting young, educated, middle class supporters. Having discontinued their high profile populist approach, when the NF moved away from their urban base, they alienated their traditional supporters without being able to appeal to a new audience. There were also extensive changes in the NF's ideological stance, which came to include support for Colonel Gaddafi, Ayatollah Khomeini, and black separatists like Louis Farrakhan[48], as part of a racialist but 'non-racist' policy. A large number of ordinary members found this difficult to accept. With a diminishing number of supporters the NF became too small to be effective and increasingly dominated by an isolated leadership. Meanwhile rival organisations like the British National Party or the NF Flag group were there to offer disaffected members a more familiar approach[49].

So it can be seen that the NF's adoption of environmental concerns was only one amongst many changes of policy, but this too was not a great success. The NF proved unable to open up a dialogue with green and other groups, their racist and authoritarian politics acting as the main deterrent. Not only did the NF's understanding of green issues remain at a rudimentary level but their sincerity was open to considerable doubt. As the Greenwave example shows, when they did set up their own green organizations, these were opportunistic, ineffective, poorly run, and with obvious links to the NF. Having adopted green ideas and convinced themselves that they were greener than anyone else, the NF made no significant attemp to translate this into policy or activity. Despite this, though the NF's attempt to adapt fascist ideology to contemporary circumstances involved a number of contradictory elements, the mixture of fascism and ecology did seem to be an effective part of the package as far as members were concerned. It also seems to have been a common element in similar attempts to construct a new style of fascism by extreme right groups in Italy, France, Germany and the USA[50].

A comparative perspective

One way to illuminate the NF's particular view of green issues is by means of a contrast with the UK Green Party. Like the NF it is a recently formed, small political party occupying a marginal position in regard to the British electoral system with comparable levels of support and membership[51]. Both Parties constitute a relatively new political force, as well as acting as pressure groups motivated by an overriding imperative: racial nationalism for the NF and protection of the environment for the Greens.

The main strands of the Green Party's ideology could be stated as ecology, democracy, social justice, and non-violence, combined with a global and holistic worldview that seeks a sustainable and improved quality of life for all[52]. Green ideology, partly influenced by the science of ecology, tends to emphasize the harmony, co-operation and interdependence of ecosystems and extend these concepts to human societies. Petra Kelly of Die Grünen has talked of the spiritual base of green politics teaching respect for all living things since all life is part of an interconnected web[53]. In the words of the Green Party, 'Green Politics is about building a new way of life .... based on respect for the planet and humility about our role on it'[54]. In the context of the current discussion it is important to note that the Green Party also states its total commitment to the eradication of racism[55].

Green politics may be influenced by ecophilosophy but it is difficult to find much material regarding this in the publications of the Green Party[56]. The Manifestos contain detailed lists of policies whilst the Party newspaper Econews is concerned with the practicalities of running a political party and furthering the cause it represents. Issues are debated but underlying beliefs are rarely examined. Instead the Green Party are oriented towards formulating responses to environmental problems and building a green society.

In the Dialectics of Nature [57], Engels observed the tendency of groups to project onto nature their own interests and preconceptions. He was specifically referring to those proponents of Darwinism who saw nature in terms of struggle. Engels saw this as nature interpreted in terms of the social philosophy of Hobbes and Malthus, then read back as eternal law. He compared this with views emphasizing harmonious co-operation within nature, making the point that the two perspectives were one-sided as nature encompassed both. Taking this observation as a starting point, a similar disparity can be identified in the conceptions of nature apparent in the ideologies of the Green Party and the NF. Another way of looking at this is in terms of the sociology of knowledge, relating the ideas held by these groups to their social location and historical experience.

Though green ideology has developed as a response to the environmental problems created by the economic and social practices of contemporary society, it also reflects the material situation of the group articulating these ideas. As many as two thirds of the environmentalists studied by Cotgrove in the late 1970s were located in the non-productive sector, with 43% employed in the caring professions. He found that this group placed an emphasis on social relationships, communal values, and participation in decision making, whilst expressing opposition to free market capitalism and unfettered economic growth. Cotgrove related this to their occupational position within a public sector coming under increasing threat [58].

More recently Rüdig 's survey of Green Party members [59] found that 44% were employed in the public sector, men and women were equally represented, most were in the 25-49 age group, 67% either had studied, or were studying for degrees, 60% lived in the south of England and more than 50% lived in rural areas or small towns. This confirms Cotgrove's findings but does not explain why this groups' concern should have crystallized around environmental issues. Williams and Eckersley have both suggested that increased social mobility, education and leisure amongst the new middle class may have increased their awareness of the problems facing modern society, including the danger of an environmental crisis [60]. Gouldner and Offe have also both noted that members of this knowledge based class see themselves as representing all sectors of the community, as well as posessing a universalistic outlook that can transcend national boundaries and take a global perspective [61]. Though it would be too reductionist to see the Greens exclusively in class terms, as these surveys show, such a focus can be revealing.

The NF represented a very different set of interests, though much less detailed information is available on this due to their reluctance to allow researchers free access to its membership. The leadership of the NF tended to be drawn from a middle class background. Of the group who took over the leadership in 1984 only one could be described as working class. The rest had all received some higher education, and whilst many of them had degrees, they tended to be self employed or work in the private sector. In Fielding's view [62] the NF's leadership and activists constituted a closely bound group, regarding themselves as an elite in posession of a fundamental truth. Their socialization into a total belief system popularly regarded as extreme and unacceptable, along with the conspiratorial form of these beliefs, inclined them to take a suspicious and combative stance towards outsiders. As a result this inner group tended to insulate themselves from the rest of the world and screen out information not conforming to their worldview.

However in order to perpetuate their beliefs the NF had to find a means to make an appeal to a wider audience. Until the early 1980s the basis for this was their call for the 'repatriation' of Britain's non-white population. Husbands' research [63], undertaken in urban areas which had shown relatively high electoral support for the NF, confirmed past impressions that those who expressed strong support for the NF, were predominantly male, aged 18-25, from the skilled and unskilled manual sector, with a substantial portion of unemployed and marginal self-employed. Thus the politics of the NF could be seen as having largely represented working class and marginalized groups living in declining inner city areas.

The new leadership, in abandoning this social base, lost their constituency without being able to build a new one. Their deliberate formulation of a new ideological position to solicit an audience amongst those sympathetic to green ideas remained an artificial construct in that this outlook had no social base outside a small group of activists. As Mannheim has suggested, attempts by a political movement to seek a 'style of thought' with which to ally itself cannot succeed if this does not relate to the social reality of an existing group [64]. Prospects of an alliance with the Greens were nothing more than a fantasy of the NF leadership.

In seeking to appeal to a green audience, the NF were constrained by their narrow understanding of what green meant. For them it largely consisted of a form of conservationist environmentalism grafted onto their nationalist and racist outlook. Though they attempted to bring in issues of lifestyle, technology and ecological perspectives, these remained subservient to strategic considerations and were interpreted in terms of the NF's core ideology. The NF seemed to conceive of green issues more in historical terms rather than as a response to contemporary environmental problems. Evidence of this can be seen in their aim to recreate a traditional and hierachical caste society of landowning peasants, craft guilds, and warriors on the model of Medieval Europe [65]. This backward looking perspective was also apparent in their use of British traditions to evoke nostalgia for the myth of England's green and pleasant land.

The NF's ideal society shows marked parallels with the German concept of Volksgemeinschaft, the soul of the people embodied in the national community. This saw the nation as an organic whole, united by hereditary descent, language and culture. Observing that the national landscape was an integral part of this whole, Mosse has emphasized the backward looking nature of these perspectives, and noted their influence on Nazi ideology [66].

However it should also be stressed that Nazi ideology was technologically modernist in outlook, and can be seen as having made an opportunistic use of the appeal of rural romanticism in celebrating the bond of 'blood and soil'[67]. The Nazis also saw biological and genetic factors as more significant than environmental ones, something they combined with Social Darwinist conceptions of nature as red in tooth and claw. Existence was portrayed as a struggle in which only the fittest survived, with the, 'aristocratic principle of Nature'[68] justifying hierachy, male dominance, inequality, and war. For the Nazis the prime site of this struggle was between the 'white races' and other ethnic groups who were portrayed as inferior or sub-human.

The NF were strongly influenced by such conceptions of nature, especially in their advocation of the moral and spiritual benefits of a return to contact with the soil. This had racial connotations since homogeneous communities were seen as more stable, racial conflict being the result of people being taken out of their natural environment[69]. The NF also used sociobiological theories to support their viewpoint in that, 'Sociobiologists have shown that racialism, far from being an anti-social aberation, is an instinct inherent in all peoples and species. They have shown that racialism is the driving force behind evolution'[70]. So for the NF this meant that, 'No National Revolutionary can afford to view the Race, Green and Morality issues as separate and unrelated. Our ideology is akin to our view of the Nation: an organic and interdependent whole'[71].

Yet this attempt to combine green issues and those of race also raised contradictions. When discussing the similarities between their position and that of the Greens the NF always made the proviso that though the Greens were correct in many respects, their worldview was incomplete because they ignored the issue of race, which the NF saw as a fundamental part of ecology. Because the Greens did not see the necessity of preserving the 'white race', which the NF saw as an endangered species, the Greens were portrayed as hypocritical for not taking ecological perspectives to their logical conclusion[72]. However, it was the NF's attempt to interpret ecology in terms of race and nation that constrained them from adopting green ideas in global terms and as a holistic package. The NF could not move beyond seeing green issues in terms of national and racial interests. Unlike the inclusive and international outlook of the Greens, the NF's politics not only involved the exclusion of large sections of the community, but also, they were unable to offer any solution to global environmental problems since their policies ended at Britain's borders.

Conclusion

On the basis of this comparison it would be difficult to categorize the NF's politics as green. Nevertheless, their attempt to adopt green concerns provides a means through which to investigate green ideology from a contrasting perspective. Despite their opportunistic and instrumental approach, in their own peculiar way the NF did attempt to come to terms with green ideas. But given the divergence of their worldview from that of the Greens, it can be seen that the NF were not speaking the same language when it came to their understanding of ecological concepts. So as well as pursuing different objectives from the Greens, it can be argued that the NF failed to comprehend green ideas in their totality because of their existing ideological outlook. Thus green ideas remained subordinate to the dominant elements in the NF's ideology: race and nation, nostalgia for Britain's pre-industrial past and the desire to create a traditional and homogeneous folk community. The NF's conception of nature also stood in direct contradiction to that of the Greens by stressing the primacy of genetic factors over environmental ones and emphasizing the struggle between species or races, with the so called 'white race' standing at the peak of a hierachy. Such a conception of nature could in part be related to the narrow set of class interests represented by the NF, their appeal being based on local and national self interest and territoriality - the politics of exclusion rather than inclusion.

This is unlike the Greens' contemporary orientation which focuses on finding solutions to problems concerning the relationship between modern society and its environment. The basis of their ideology is a holistic view of nature which sees society in terms of the interdependent relationship of the inhabitants of an ecosystem and the environment. The Greens, though based in the new middle class, have sought to appeal to all groups and interests, conceiving of environmental problems in global terms well as local ones. This conception of nature would seem to reflect a very different social outlook. Green ideology can be seen as a distinct package of ideas and practices arising from a particular group's perception of an environmental crisis facing modern society. Since the Green movement were the group within which this outlook developed into a recognizable worldview, they remain the measure with which other expressions of environmental concern must be compared.

This is not to say that green ideas have a necessary location, or that any one group has a monopoly of concern. The rapid spread of environmental awareness throughout society would suggest an interest in these matters extending far beyond the originating group. All sections of society experience environmental problems, though those expressing the most concern are not necessarily the most severely affected. The Green movement have certainly acted as agents for change in defining and popularizing these concerns, though other groups, as they become aware of environmental risks are bound to formulate their own responses to them. Green ideas are bound to take diverse forms as they are adapted to different group situations and interests. The motive for such adoptions may be opportunistic or amount to a form of co-option, but any credible claim to represent green concerns will also involve some attempt to provide solutions to environmental problems. The appropriation and articulation of green ideas by different groups needs to be evaluated in terms of their social location, the interests they represent, what relationship exists between green ideas and other elements of their ideology, and what form their responses to environmental problems will take, for it can be seen that some approaches are more desirable and more green than others.

Acknowledgements

I would like to thank both the Wiener Library and the Institute of Jewish Affairs for their assistance in providing access to source material. Thanks are also due to Ruth Levitas for her comments on an earlier draft of this paper.

Notes and references

(1) M Billig, *Fascists*, London, Academic Press, 1978, see Ch 6.
(2) On the NF and Nazi ideology, see Billig *ibid*. and S Taylor, *The National Front in English Politics*, London, MacMillan, 1982, pp64-81.
(3) See R Thurlow, *Fascism in Britain*, Oxford, Basil Blackwell, 1987, pp285-6, Christopher Husbands, *Extreme Right Wing Politics in Great Britain: The Recent Marginalization of the National Front*, West European Politics, Vol 11 No 2, April 1988, pp61-79, and Taylor *op cit*.
(4) In 1980 John Tyndall was ousted from the NF's leadership, see *Searchlight* 61, July 1980, pp4-7. He went on to set up the rival British National Party.
(5) Terza Posizione was associated with Nuclei Armati Rivoluzionari, a group accused of acts of terrorism. Several of those who fled to Britain were sentenced in their absence by Italian courts: see *Searchlight,* From Ballots to Bombs, London, Searchlight Publishing Ltd., 1989, p7; *Nationalism Today* 40, pp14-15; and *NT* 46, July 1989, pp14-16.
(6) The Italian fugitives and the NF activists held joint seminars and together produced an occasional magazine called Rising, see *Searchlight,* From Ballots to Bombs, *op cit*., p8
(7) This was done through two new publications, *Nationalism Today* (NT) and *New Nation* (NN). *National Front News* and *Bulldog* continued with the NF's familiar racist and populist stance.
(8) *NT* 7, p20.
(9) *NT* 3, p13.
(10) *NT* 9, pp12-13 and *NT* 26, Jan 1985, pp10-11.
(11) *NT* 14, p20.
(12) *NT* 8, p6.
(13) A Healthy Race, A Clean Environment, *NT* 12, p5.
(14) The Urban Animal, *NT* 19, p13.
(15) *NT* 25, Nov 1984, p14.
(16) New Technology: Master or Servant, *NN* 5, pp18-19.
(17) Steps Towards a New Society, *NN* 5, p14. See also Towards a Nationalist Community, *NT* 32, p15.
(18) More positive examples of utopias with an environmental dimension are contained in: E Callenbach, *Ecotopia*, London, Pluto Press, 1978;

A Gorz, *Ecology as Politics*, Boston, South End Press, 1980; R Bahro, *The Alternative*, London, Verso, 1978 and *Building the Green Movement*, London, Heretic, 1986. By way of a contrast, during this period the NF were distributing *The Turner Diaries* by A Macdonald (W Pierce), National Alliance Books, Washington DC, 1978. This was a singularly unpleasant racist utopia in which a group of white supremacists take over the world, wiping out all 'non-Aryans' and rendering a large part of the planet uninhabitable through the use of nuclear and biological weapons, see review in NT 4, p13.

(19) An example of this was Gandalf Graphics, a business in Suffolk run by NF members, which received a grant from the Manpower Services Commission. Under MSC rules their funds cannot be used for political purposes but Gandalf Graphics supplied work to the East Suffolk Press which printed the NF's publications. This prompted an MSC investigation, see Searchlight No 126, Dec 1985, p8.

(20) After their 1983 electoral failure this group got rid of another Chairman, leaving them in control of the NF's Directorate, see *Searchlight* 104, Feb 1984, pp10-11.

(21) *NT* 20, April 1984, p2.

(22) See *NT* from September 1984 onwards. Issues of race were largely absent from these features.

(23) In the 1986 split the group popularly known as the Political Soldiers held the Directorate. Those forced out set up a second NF known as the Flag group which returned to former strategies, see Searchlight 132, June 1986, pp2-3.

(24) *NT* 37, Mar 86, p6.

(25) The cover of *NF News* 99 (late 1987) had a picture of a black clenched fist with the slogan 'Smash Racism'. Many NF Branches just sent their copies back, see *Searchlight, From Ballots to Bombs*, pp34-35 *op cit.*

(26) For the NF's attitude to the Third World see *Why Band Aid Will Kill Blacks*, NT 34, Oct 1985, p9 and *Colonialism - Racial Disaster* NT 22, p13.

(27) In September and October, 1988, Margaret Thatcher made two key speeches declaring her commitment to protecting the environment, see Margaret Thatcher, *Our Threatened Environment: The Conservative Response*, London, Conservative Political Centre, 1990. This event generated high levels of political, public and media interest in green issues.

(28) *NFN* 125, Feb 1990.

(29) On Evola see *Rising* No 2, 1981 and No 3, 1982. On Codrianou see *Yesterday and Tomorrow*, Rising Press, 1983. On Darrß also see *Rising* No 2. On the Strassers see NT 18, p12.

(30) See Our Radical Roots, *NT* 19, pp8-9 and The Roots of British Nationalism, *NT* 26, Jan 1985, pp14-15.

(31) See *Tolkien Ring Bearer for Racial Nationalism*, New Nation 6, 1984,

pp4-5, review of Lord of the Rings, *NT* 27, p16, and *Searchlight* 122, Aug 1985, p8.

(32) The symbol of a stylized Celtic cross appeared in Rising in 1983 and by 1984 it had begun to appear in the pages of NT. After the split with the NF Flag group in late 1986, the Union Flag on the masthead of NF News was replaced with the cross within a circle symbol. The image of a more traditionally decorated Celtic cross was also frequently used for illustrating articles and publicity material.

(33) See for example *The Times*, 20.10.84. A more critical perspective was given by Sarah Benton & Rob Edwards, The 'Greening' of the National Front, *New Statesman*, Vol 108, No 2797, 26.10.1984, pp16-17

(34) Step Up the Propaganda War, *NT* 32, Aug 85, p23.

(35) See *NT* 44, Jan 1989, p23 and *On the Attack*, Burning Books, Croydon, 1987, an NF manual on infiltration and intelligence gathering.

(36) The Tide Has Turned, *NT* 18, p9.

(37) A Society in Revolt, *NN* 5, Spring 1984, pp8-9.

(38) As the NF had tended to do in the past. Even after 1980 the NF's attitude towards the Greens had been one of hostility. During the run-up to the 1981 GLC elections, the home of Paul Ekins, the General Secretary of the Ecology Party, was attacked, with a warning not to stand against the NF's candidate. Soon afterwards Ekins received a leaflet from the NF in which they claimed to be the Party of the environment, see *The Guardian* 24.2.81.

(39) In the 1989 European Elections the Green Party gained 14.9% of the total UK vote, the highest proportion of a national vote achieved by any European Green Party.

(40) The Green March, *NT* 46, July 1989, pp12-13.

(41) This was another development following Margaret Thatcher's speeches on the environment. See the interview with Greenwave's co-ordinator in *Whole Life Magazine*, Aug/Sept 1989, p42.

(42) The name Greenwave had originally been intended for a new environmental group being founded by Peter Wilkinson of Greenpeace. When the NF appropriated the name of his organization Wilkinson decided not to sue as he thought it would give the NF a publicity coup, see *The Guardian* 6.1.90. and *Searchlight* 166, April 1989 p3.

(43) Green Dawn, Sept 1989.

(44) See *Time Out* 26.4.89 and *Searchlight* 168, June 1989, p14.

(45) ibid.

(46) These figures on NF membership are based on estimates given by Thurlow op cit., and Husbands 1988 op cit. See also *Searchlight* 127, Jan 1986, p5, *Searchlight* 163, Jan 1989, pp9-11.

(47) *Guardian* 27.4.91.

(48) On support for Gaddafi, Khomeini, and Farrakhan see *NF News* 103,

March 1988. Also see *Searchlight* from Ballots to Bombs *op cit.*

(49) On reasons for failure see Larry O' Hara, Bibles, Business Cards and Jackboots, *New Socialist* 68, Aug/Sep 1990, pp29-31 and C Husbands, 1988 op cit.

(50) On attempts to adopt green issues by the extreme right see the following: on Italy see *the Economist* 11.10.80, p56 and an article on MSI candidate Alessandra Mussolini in *Weekend Guardian* 22.2.92, p6; on France, see report on the National Front's conference on ecology, *Financial Times* 2.4.90 and the description of the attempt by Nouvelle Résistance, a Third Positionist group, to infiltrate Ecolo-J, *Searchlight* 211, Jan 1993; on Germany see report on Republikaner policy, *Searchlight* 172, Oct, 1989 p12 and account of the extreme right connections of the Ecological Democratic Party, *Searchlight* 189, Mar 1991; on the USA see article on White Aryan Resistance, *Searchlight* 171, Sept 1989, p17.

(51) On Green Party membership and electoral performance, see section on UK in Sara Parkin, *Green Parties: An International Guide*, Heretic Books, 1989 and W R dig and P Lowe, The Withered 'Greening' of British Politics, *Political Studies* 34, 1986, pp262-284.

(52) On the ideology of the Green Party see *Manifesto For a Sustainable Society*, the Green Party and Jonathon Porritt, *Seeing Green*, Oxford, Blackwell, 1984. For a comparison with Die Grünen see *Programme of the German Green Party* with a preface by Jonathon Porritt, London, Heretic, 1983.

(53) Petra Kelly, Towards a Green Europe and a Green World in Felix Dodds (ed), *Into the 21st Century*, London, Greenprint, 1988, p 109.

(54) Green Party, General Election Manifesto, 1987, p 1.

(55) *ibid.*, p14.

(56) Something also noted by Andrew Dobson in *Green Political Thought,* London, Unwin Hyman, 1990 p68.

(57) Frederick Engels, *Dialectics of Nature*, Moscow, Foreign Languages Publishing House, 1954, p404. See also Raymond Williams' essays, *Ideas of Nature and Social Darwinism*, in *Problems of Materialism and Culture*, London, Verso, 1980.

(58) Stephen Cotgrove, *Catastrophe or Cornucopia*, Chichester, John Wiley and Sons, 1982.

(59) W Rüdig , L Bennie and M Franklin, *Green Party Members: A Profile*, Glasgow, Delta Publishers, 1991.

(60) Raymond Williams, *Towards 2000*, London, Chatto and Windus, 1983 and Robyn Eckersley, *Green Politics and the New Class*, Political Studies 37, 1989, pp205-223.

(61) See Alvin Gouldner, *The Future of the Intellectuals and the Rise of the New Class*, London, Macmillan, 1979 and Claus Offe, New Social Movements: Challenging the Boundaries, *Social Research* 52, No 4, Winter 1985, pp817-868.

(62) Nigel Fielding, *The National Front*, London, Routlege and Kegan

Paul, 1981.

(63) Christopher Husbands, *Racial Exclusionism and the City: The Urban Support of the NF*, London, Allen and Unwin, 1983.

(64) Karl Mannheim, Competition as a Cultural Phenomenon in *Essays on the Sociology of Knowledge*, London, Routledge and Kegan Paul, 1952, pp 209-210. See also Ruth Levitas, The Social Location of Ideas, *The Sociological Review*, Vol 24 No 3, Aug 1976, pp545-557.

(65) On the NF's Medieval outlook see *Rising* No 2, 1981 and No 3, 1982. On Belloc, Chesterton and Distributism, see *NT* 20, p9; *NT* 33, Sept 1985, p2; and *NT* 45, April 1989, pp14-16.

(66) G Mosse, *The Crisis of German Ideology*, London, Wiedenfeld and Nicholson, 1966, pp 4-5, also Ch 1. A more problematic account is given in Anna Bramwell *Ecology in the 20th Century*, London, Yale University Press, 1989, and her earlier book, *Blood and Soil: Walther Darrß and Hitler's 'Green Party'*, Bourne End, The Kensal Press, 1985. Using a rather limited conception of what constitutes green politics, Bramwell implies that Nazi Germany was the first 'green' state, though she provides little in the way of convincing evidence to substantiate this claim.

(67) This commitment to rural romanticism could be seen as more rhetorical than actual, see J Herf *Reactionary Modernism*, Cambridge, Cambridge University Press, 1984, especially Ch 8. Z Bauman, in *Modernity and the Holocaust*, Cambridge, Polity, 1991 also argues that though the Nazis made use of anti-modernist sentiments, their racist ideology, policies and practices, were a product of modernity. Similar arguments made at the time of the Nazis rise to power are contained in Ernst Bloch, *Heritage of Our Times*, Cambridge, Polity, 1989 and Walter Benjamin, *Theories of German Fascism*, New German Critique 17, Spring 1979, pp120-128.

(68) See Adolf Hitler, *Mein Kampf*, London, Hutchinson, 1969, p348 and Rick Wilford, Fascism in R Eccleshall et al, *Political Ideologies*, London, Hutchinson, 1984, p232.

(69) See NF *Statement of Policy*, 1983 and *Introduction to the Movement*, NF, 1989.

(70) NF *Statement of Policy op cit.*

(71) *NT* 45, April 1989, p23.

(72) See The Green March, *NT* 46, July 1989, pp12-13.

2 The mixed fortunes of the British Green Party 1989-1992

Wolfgang Rüdig, Lynn Bennie and Mark Franklin

Introduction*

The roots of Europe's first Green Party were planted in February of 1973, when 'People' was born. Since that time the Party has experienced two changes of name but, unlike some of its younger European cousins, little success electorally. For most of its two decades of existence the British Green Party[1] has never attracted more than 1-2% of the vote. There was one brief period however when all of this changed. Between 1989 and 1990 the Green Party suddenly emerged as a prominent political force in British national politics. The Party polled 14.9% in the 1989 European elections, and more than doubled its membership in the ensuing year to nearly 20,000 members. However, by 1993 a very different political reality faced the British Greens. The Party's poll standing had returned to that of its earlier years, and the 1992 General Election had proved nothing less than dismal. In that election, the Greens polled an average of 1.3% in the constituencies it contested and 0.5% of the national vote. Party membership reflected this decline; well over half of the Green Party members who had joined in the boom of 1989-90 had since left, leaving it with a membership of below 10,000. In short, the British Green Party had returned to the gloomy political wilderness in which it barely survived prior to 1989 (see Figure 1).

**Revised version of paper presented at the IRNES conference Perspectives on the Environment: Research and Action into the 1990s, Leeds University Conference Centre, 14 and 15 September 1992.*

Figure 1: Poll Standing and Membership Development of the British Green Party, 1988-1992[2]

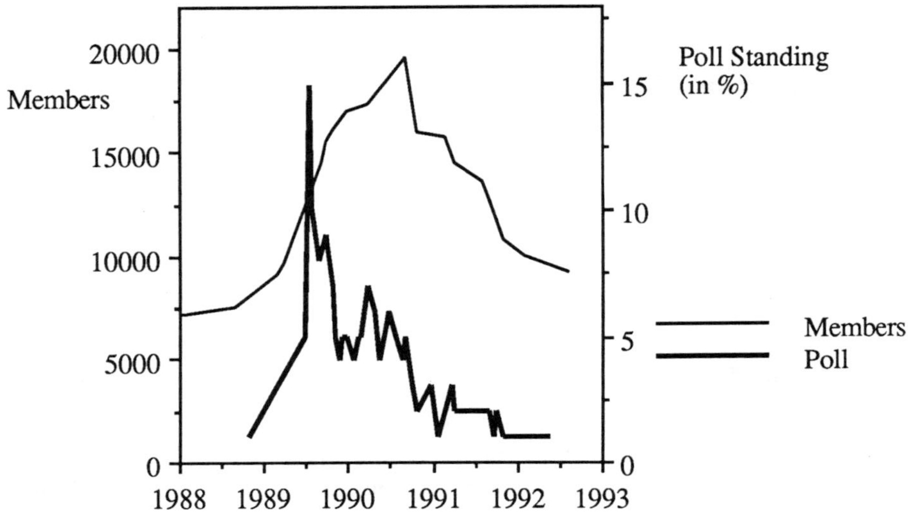

The fluctuating pattern of green party politics in Britain contrasts distinctly with the broad trends and characteristics of other European green parties. In fact, the experience of the British Greens appears remarkably out of sync with those of their European counterparts. Before 1989, it seemed safe to assume that Britain could be classified as belonging to those countries in which a green party could not be expected to play an important role (Rüdig and Lowe 1986). In 1993, that assessment could fairly be made once again. What happened in between to make it appear as if the hour of the Greens had finally arrived? And why did this period of political excitement turn out to be so brief? These are the questions we seek to answer in this paper.

To understand fully the British Green Party's development over these interesting years, we need to analyse the reasons for both the up-swing in the Party's fortunes and the down-swing. The paper is therefore separated into two main sections. We first attempt to make sense of the up-swing period when the Party experienced its only taste of real electoral success. We look at the general political environment of the time, why people voted for the Green Party in the European Elections, and why people joined the Party in 1989. We then turn our attentions to the other side of the coin by providing an overview of the down-swing period. How did the political scene differ at this time? Why did the Party's '89 electorate largely fail to vote Green again in the 1992 General Election, and why did members leave the Party in such large numbers after 1990?

The sources on which our analysis is based are three-fold. First, we observed the political context from 1989 to the present time, attending Green

Party conferences, monitoring media coverage and the general political environment.

Second, for our analysis of Green voting behaviour, we rely on data collected by the 1989 European Election Study. These data provide information about actual green voters in the 1989 European Elections, and about the future intentions of 1989 green voters.[3]

Third, for our analysis of Party members, we draw on a panel study of Green Party members conducted in 1990 and 1991.[4] In the first wave, a random sample of members of the Green Party was mailed a detailed questionnaire in November 1990. About 8,600 questionnaires were mailed,[5] 4,357 usable questionnaires were returned, a net response rate of 51%. For the second wave, 4,000 respondents to the first wave were contacted again.[6] A second questionnaire was mailed to this group in November 1991. A total of 2,651 usable questionnaires were received, a response rate of 66%.[7]

1989: the European election success

Why did the Green Party suddenly burst on to the political scene in 1989? How did they achieve their unprecedented success in the elections to the European Parliament in that year? We begin by looking at the political context which provided the foundation for the rise of the Greens (cf also Frankland 1990; Rootes 1991).

The environment suddenly climbed to the top of the political agenda in 1988. An unprecedented stream of environmental issues were brought to the attention of the public. Undoubtedly the salience of environmental issues had reached an unprecendented level by 1989. The Chernobyl accident had occurred in 1986 and although there had been a somewhat muted reaction to this in Britain (Rüdig, 1990) it provided a background for the environmental issues which were about to be brought to the fore. The environmental hazards of the Sellafield complex and electricity privatisation, including its nuclear component, attracted public concern. The Government's plans for the privatisation of water met with increased unpopularity (cf. Richardson et al. 1992). Other issues which prompted public concern included the odyssee of the ship Karen B, laden with toxic waste, and the dying of the seals in the North Sea. In Britain, food safety, specifically eggs, suddenly became an issue, resulting in the resignation of a high profile health minister. Finally, an important event which helped bring the environment to the top of the agenda in Britain was the Royal Society speech made by the then Prime Minister Margaret Thatcher in September 1988. Using typically green expressions and slogans Mrs Thatcher embraced the 'greenhouse effect' as the centre-piece of the Government's new environmental initiative (Flynn and Low, 1992).

As environmental issues gained credence amongst the electorate, 1988 and 1989 saw all of the established parties scrambling for a suitable green image. The Conservatives had a very poor environmental image at home and in Europe (note Britain's label 'dirty man of Europe'). Having previously rejected green arguments and programmes, the Conservatives were now faced with a leader who had suddenly sensitised the Conservative electorate to environmental concerns.

The Labour Party was faced with a different set of problems. The Party had in previous years adopted a fairly radical environmental programme by committing itself to unilateral nuclear disarmament and a non-nuclear energy policy. However, reeling from election defeats in 1983 and 1987, the Labour leadership began modernising the Party in an attempt to win back the 'middle of the road' voter. This involved a watering down many of its radical policies including its stance on unilateral nuclear disarmament. Thus, supporters of radical ecological and anti-nuclear positions lost out in the new look Labour Party.

The Liberal Party had been the most environmentally conscious of all the parties. In 1979 its manifesto looked very 'green' and it was the first party to reject nuclear energy (cf. Mez and Rüdig 1979). However, the wider swing towards environmental awareness came at a bad time for the Liberals. Their alliance with the Social Democratic Party (SDP) had failed to make the anticipated breakthrough and the merger of the parties following the 1987 election proved less than amicable. The turmoil of the merger made them very unpopular. In addition, the SDP's policies on civil and military nuclear issues led to fears amongst 'green' Liberals that the green edge of their platform would have to be compromised. All in all, as the European Elections approached, the established British political parties all displayed severe environmental weak spots.

The European Elections provided those disaffected supporters of the major parties with the perfect opportunity to register a protest vote. European Elections are commonly referred to as 'second order elections' (Reif 1984). When no government is elected, interest levels tend to be very low, and turn-out is expected to be much lower than for 'first order' elections, such as national parliamentary elections. In this context, it is predicted that voters will be more willing to vote for small parties.

However, the Greens could only attract these votes because they were seen as a new force which deserved support. The Green Party had come a long way from its rather humble beginning as 'People'. Having established itself as Britain's most important minor party, it was able to mount a very respectable campaign in 1989. The Party mobilised the resources to field candidates in all 78 constituencies, for the first time ever. This gave the Party welcome publicity and established it as 'up-and-coming'. National visibility was enhanced by a high-profile national election broadcast, produced with the help of fashion photographer David Bailey.

In this way, the Green Party was able to grasp the opportunity that was presented to it in 1989. It attained enough visibility and credibility for it to be taken seriously by many voters for the first time. Who, then, were the voters who turned to the Party?

Our analysis of green voting (Franklin and Rüdig 1992a, b) revealed that of all demographic characteristics education is the main predictor of British green voting in 1989. British green voters also see themselves and the Green

Party as slightly more to the left than others, and voters see the Party as more to the left than themselves.

The variable we found to be most clearly associated with green voting was the importance of the environment as a political issue. 53% of 1989 green voters regarded the environment as one of the top three political issues, compared with 30% of the electorate. Evidently, the rise of the environment as a political issue in 1988 and 1989 did influence the green vote.

An examination of previous voting behaviour shows how broadly based green voting was in 1989. Table 1 reveals that support for the Green Party came from previous supporters of all the parties, not just the centre-left parties. Most prominent are the previous Alliance voters and least prominent are the previous Labour supporters.

Table 1: Previous Voting Behaviour of 1989 Green Voters

Party Supported in 1987 General Election	1989 Green Voters %	All 1989 Voters %
Conservative	24.7	38.8
Labour	18.5	32.7
Alliance	27.1	14.8
Green Party	7.4	1.2
Others	2.5	2.1
Abstention, Don't Know	19.7	10.4

The characterisitcs of the new Green Party members of 1989 are contrasted with those of Green voters and the general population in Table 2. Members appear relatively young, are more highly educated, have professional occupations and place themselves left of centre. If we compare new members with 1989 voters, we find that members are drawn from a fairly narrowly defined group of people, while voters came from a broad section of the population. Comparing new and old Green Party members (Rudig et al. 1991), there is hardly any difference in their profile in terms of socio-demographic background as well as their past activities. Past membership of other political parties, social movement activities, and environmental and peace group membership are broadly the same for old and new members. For instance, less than 30% had been a member of another political party; half of these had been members of the Labour Party.

Table 2: Characteristics of New Green Party Members 1989 and Green Voters 1989

Variable		1989 New Green Party Members (N=1349)	1989 Green Voters (N=81)	Population (N=957)
Age	18-24	11.7	11.1	13.1
	25-34	33.7	27.2	20.0
	35-44	27.3	24.7	19.4
	45-59	15.8	17.3	18.9
	60 and older	11.6	19.8	28.5
Education (Age left full-time education)	less than 15	10.6	24.7	38.8
	16-17	14.8	32.1	35.8
	18-20	14.5	19.8	11.4
	21 or more	60.1	23.5	14.0
Occupation	Manual	7.2	22.2	25.3
	White Collar/ Professional	60.0	42.0	27.3
	Retired	7.7	12.3	21.0
	Housewife	7.4	9.9	18.0
	Student	10.4	6.2	4.2
	Unemployed	7.2	3.7	3.3
Left-Right Orientation (Mean Score) 1 Left- 10 Right)	Personal	4.2	5.3	5.7
	Green Party	4.1	4.6	4.4.

Hence, the 1989 upsurge in Green Party membership did not represent a mould-breaking period in terms of attracting new sections of the population to the Party. There is no evidence of a large-scale influx of disgruntled green activists from the Labour Party or the Liberal Democrats, for example. The general political conditions, e.g. the electoral system, facing the Greens remained similar to the pre-1989 era. As before, the British political process ensured that small parties had practically no chance of establishing a permanent foothold. Consequently, it could have been expected that membership commitment patterns and the duration of membership would not change substantially. In the past the average length of membership had been 2 years but the Party was relatively good at attracting new members. After 1989, the drop-out rate of members was quite similar to those experiences in previous years but the Party found it more difficult to attract new members.

1992: The General Election fiasco

By the end of 1990 the fortunes of the British Green Party began to look less favourable. The Party's opinion poll standing started to decline sharply, its members began to leave in larger numbers and they were not replaced by new members. What had happened?

The political situation facing the Greens in the run-up to the General Election of April 1992 was very different from 1989. To begin with, the general economic situation had changed from prosperous boom to deep recession. The political agenda had changed with concerns about the environment being pushed to the margins by traditional economic concerns about unemployment, interest rates etc. Media exposure of environmental matters had reached saturation point by 1990. All in all, environmental issues began to lose their salience. These facts combined meant that the environment had little role to play in the election campaign.

The political parties in Britain had also undergone some change. The Liberal Democrats had put the merger fiasco behind them and were well on their way to recovering their place as the third party force in British politics. There was also real optimism in the Liberal Democrat ranks that the Party might achieve the balance of power at the election. With this prospect in sight, Liberal Democrat supporters had good reasons to vote for 'their' party.

The Labour Party, for the first time since 1979, seemed to have a real chance of an out-right victory in the election. The Conservatives at some points in the campaign seemed destined to lose many of their seats. The closeness of the election meant that supporters of these parties were motivated to stick with them rather than turn to another party.

In addition to the different structure of the political agenda, this was now a 'first order' election where experiments are shunned, the 'lost vote' argument is more pervasive and the turn-out is much higher. The higher number of constituencies made it far more difficult for the Party to field a full slate of candidates. The majority voting system could exert its usual detrimental effects on small parties. In fact, even if the political agenda had been substantially more pro-environmental shaped, the electoral system would have made it extremely difficult for the Greens to have made any significant impact.

All in all, it is thus not difficult to see why the Greens were squeezed out of the competition. Was the Green Party in any position to counteract these trends? It seems not. Apart from the unfavourable external conditions which were beyond the Party's control, the internal development of the Party did not help. The enthusiasm of 1989 had long since worn off and the Party found itself consumed with internal disputes over its internal structure. A major internal reform programme, called 'Green 2000', was heatedly discussed and finally adopted in September 1991. The new structure streamlined the internal organisation by introducing an executive council and two principal speakers in the hope it would give the Party a more coherent image and a stronger national decision centre. The debate over this initiative in most of the media had a rather negative slan, focussing on 'splits'. Financial problems were one more hindrance. The Party had expanded its activities in the days of its success but when new members failed to join in sufficient numbers and renewal rates started to drop, the financial basis of the Party became very weak. Beset by these problems the Party was ill-prepared for the election in April 1992. The Party put forward 253 candidates, up considerably from 133 in 1987, but this compared unfavourably with the European Election when they fielded a full list. The media no longer looked to the Greens for novelty interest; instead they were attracted to other new parties like the 'Natural Law Party'. In the event the Greens faired poorly in the election. The result of 1.3% in the constituencies they contested was slightly down on the 1.4% they had polled in 1987.

Looking back at the characteristics of the 1989 vote, this may, however, not have been very surprising. The 1989 green voter was not very strongly committed to the Green Party. Of those who voted green in 1989, 42% said that they would vote for the Green Party again at the next General Election. Only 28.4% said that they felt 'close' to the Green Party. The level of commitment demonstrated is very much lower than in other European countries. For example, 86% of 1989 green voters in Luxembourg, 76.3% in Belgium and 73.3% in Germany said they would vote green again at the next general election. And 62.8% of 1989 green voters in Germany, 62.1% in Luxembourg and 55.9% in Belgium, stated that they felt 'close' to the Greens. In total, those who voted for the British Greens in 1989, felt close to them, and also intended to vote green at future national elections only make up 1.9% of the total sample. Other Europe green parties enjoy a slightly higher level of core support: 5.5% in Belgium, 5.3% in Luxembourg, and 4.2% in Germany.

In summary, the evidence on green voting indicates a number of factors. Support for the Green Party seems to be somewhat volatile. There is quite a large pool of potential support in the British electorate but this support is dependent on the context of the election. In other words, the Party has not managed to build up any sizeable core support yet. In the adverse circumstances of 1992, the result was actually quite respectable.

While the loss of voters compared with 1989 was thus more or less predictable, the severe loss of members constitutes perhaps more of a surprise. The Green Party lost half of its membership in the period between the end of 1990 and the middle of 1992. The new members who joined in 1989 were largely responsible for such a rapid fall in membership numbers. By November of 1991, 40% of 1989 members who responded to our second questionnaire had already ceased to be members or were about to withdraw from the Party.

In our attempt to discover why so many were leaving the Green Party we first of all asked our respondents why they thought the Green Party had declined since 1989. The options we offered respondents were evenly split between internal reasons (i.e. they could be influenced by the Green Party) and external reasons (i.e. they are beyond the influence of the Green Party).

Internal reasons included the lack of a single leader, and internal bickering in the Party. External reasons involved the evils of Britain's political system and media and the fact that there had been few big environmental disasters since 1989. In general members blame internal reasons for the decline of the Party.

When we look in greater detail at those who are leaving the Party we find leavers blame internal factors to an even greater extent. In fact we found there to be a positive correlation between leaving and blaming internal reasons ($r=.160$, $p \leq 0.001$) and a negative correlation to blaming external reasons ($r=-.148$, $p \leq 0.001$). Activists in the Party show a different tendency. They tend to blame external factors more readily. Why should this be so? It could be that activists have a belief in their own ability to change things within the Party, thus their reason for activity. It is only external circumstances that activists have no control over. The biggest difference between leavers and activists exists on the question of a single leader. Activists are less enthusiastic about the idea than leavers, and there are correlations of identical size ($r=.117$, $p \leq 0.001$) but in opposite directions between this variable and leaving and activism respectively. We went on to examine the reasons given for leaving the Green Party. Specifically, we asked our respondents why they had left or were thinking of leaving the Party.[8] The most important reason was 'lack of proper leadership', considered to be at least an important reason to leave by 75%, and the decisive reason for 27%. Closely following this was a sense of demoralisation brought on by the Party's ineffectiveness in the present political system. 73.5% considered this to be at least 'important' for their decision to leave, while for 28% it was the decisive reason. The third most popular reason for leaving is the feeling that 'my money is more effectively spent in support of pressure groups such as Greenpeace and Friends of the Earth'. 65% give this as the reason which was at least important, 26% considered it 'decisive'. Fourth and fifth place is taken by people who failed to resubscribe due to a lack of money or for personal reasons (Decisive 17% and 14% respectively). Are Green leavers likely to leave because they are attracted to another party? It appears not: only 12% give this as the decisive reason for leaving. Of all leavers, 5% have joined another party, and 12% are considering joining another party. Therefore, for more than four fifths of Green leavers, the attractions of other parties are not a consideration. For those who have joined or are considering joining another party, the Liberal Democrats are the most popular.

From our review of reasons for dropping out of the Green Party a focus on political effectiveness thus emerges as the most important factor. Given that leavers blame internal workings of the Party for its decline and demand stronger leadership, one might have expected the Green 2000 initiative to have stemmed the leaving flow. We asked our respondents what they thought of the reforms introduced by Green 2000.[9] In total, 68% thought these reforms would have a beneficial effect, 32% said they would make no difference, and only 4% thought they would have a bad effect. Perhaps surprisingly, when we compare leavers and stayers, there is hardly any difference (See Table 3).

Table 3: Assessment of Effect of Green 2000 and Decision to Leave/Renew Subscription

Green 2000	Already Left	Probably Will Leave	Probably Will Renew	Definitely Will Renew
Beneficial Effect	64.8	60.6	68.6	70.6
No Difference	32.2	32.1	27.7	25.5
Bad Effect	3.3	7.3	3.7	4.0
	N=432	N=137	N=462	N=860

If leavers believe that Green 2000 will be beneficial for the Party, why, then, do they still choose to leave? There are some indications that the reforms came too late. For example, of those leavers making up their mind in 1991, 80% had already decided to leave the Party before September 1991, i.e. before the conference approving Green 2000 took place.

However, can we assume that internal changes at an earlier stage in Green Party development would have stopped the flood of leavers? It seems not when we consider that leavers actually have a very limited knowledge of internal Party affairs. For example, of those who had left or decided to leave by Septembers 1991, only 43% had actually heard of Green 2000.

These trends relate to the sources from which Green Party members obtain their information on the Party. When asked a question on the main sources of information about the Green Party, leavers had a clear preference for the national media, television, and radio. They showed little interest in Party publications and newsletters. On the other hand, those maintaining their membership received most of their information from the Party's own publications and newsletters, and from informal sources within the Party. Consequently, it seems that leavers are least likely to be influenced by internal changes in the Party. What matters to the leaver is the Party's general political effectiveness, and this type of effectiveness is measured in very traditional ways. The preference for one national leader, for instance, suggests a very conventional political outlook. The passive members like to see their Party's representatives 'doing something' at the national level. If levels of media reporting are low, then the effect on the perception of Party effectiveness for these passive members is devastating. These results present the Green Party with some problems. Even though the organisation has been restructured in an attempt to increase the profile of the Party, if the media is not interested in the Party and does not want to talk to the two principal speakers, then passive members will still become demoralised because of the lack of information available to them from their preferred sources.

While the Green Party must be concerned about the large number of leavers, it also has to deal with the preferences of its activists, the life-blood of the Party. Activists are more skeptical about the Green 2000 initiative. Of those who consider themselves 'extremely active', around 5%, 48% thought

Green 2000 would have a positive effect, 39% thought it would make no difference, and 14% thought it would be bad for the Party.[10] Opposition and scepticism increase with rising levels of activity. However, activist disenchantment with the reforms has not led the activist element to leave the Party. Overall, activists who are against or lukewarm about Green 2000 are less likely to leave than passive members who thought Green 2000 would be beneficial.

The Green Party currently seems to be riding on the horns of a dilemma. It needs to be able to please both ordinary, fee-paying members and activists. Ordinary passive members look for Party speakers 'doing something' with a high media profile. This is something the leadership of the Party has aspired to but has found difficult. Even with an internal structure like Green 2000, it cannot guarantee media exposure at national level. Green 2000 on its own was thus unable to stem the tide of members leaving, as most Green Party leavers are passive and pay most attention to national level politcs.

At the same time, Green 2000 upset a minority of activists whose activity level is extremely high. Amongst activists, Green 2000 opponents and sceptics reach 50%. This group has not reacted to Green 2000 by leaving but these data suggest a continuously high potential for internal Party strife over issues of internal Party organisation.

Conclusions

The success experienced by the Green Party in 1989 was a clear reflection of rising environmental concern. All three major parties had failed to respond effectively to this concern. The result was that many voters chose to punish their normally favoured parties in the 1989 European Elections. The Greens themselves helped encourage the voters to send an environmental message to the other parties by presenting a unified, 'up-and-coming' image.

A completely different story unfolded at the 1992 General Election. The recession put paid to most green concerns, and the traditional economic issues once again took over.

We found that the green voters of 1989 came from very diverse socio-demographic backgrounds. In contrast, the Green Party members mobilised at the time of the European Elections proved to be of the same type of background as members the Party had attracted previously. The green message has not broken new ground in terms of membership support.

It is, of course, crucial that the Party is able to maintain its current membership. Unfortunately, it has not been able to hold onto the influx of 1989 members. Those leaving the Party tend to blame the decline of the Party on internal factors. What bonds the passive leavers to their Party is the perceived political effectiveness of the Party at the national level, where the media spot-light falls. When the media loses interest and chooses not to highlight the Green Party, the passive members themselves lose interest and conclude that the Party is 'going nowhere'.

In the final analysis, we do not see what the Green Party could have done to avoid its recent decline. The external problems face by the Party were formidable. Unlike other countries where success in 'second order' elections provided green parties with a spring-board for success in national elections, the British political system simply did not allow such a transition. The British Greens are not about to become extinct however. The Party has proved itself as a small but enduring force in British politics. It has existed on the political fringe for two decades. Today, a core support of long-term, ideologically committed activists continues its existence irrespective of happenings in the wider political world. This bed-rock gives the Party the opportunity to wait for another appropriate political context to present itself. Greens regard this as inevitable, because the fundamental problems confronted by green politics remain unsolved. Therein lies the paradox: For the Green Party to experience better times it must wait for environmental problems to worsen.

Footnotes

1. We use the term "British Green Party" throughout the paper for reasons of simplicity. In fact, there are two separate green parties in the UK: the Green Party of England, Wales and Northern Ireland; and the Scottish Green Party. All references to election results, membership development and opinion poll standings refer to both parties; all data on party members reported in the paper only refers to members of the Green Party of England, Wales and Northern Ireland.
2. Sources: For membership figures, Green Party office, London; for poll standing, *The Observer*, 1988-1992, plus selected NOP polls. The 15% mark for June 1989 refers to the European Election result itself.
3. The European Elections Study 1989 (EES 1989) is a joint effort of Western European social scientists to take advantage of the elections for the European Parliament (EP) held simultaneously in all European Community (EC) countries in June 1989, in order to engage in cross-nationally comparative electoral research. In the present paper we focus upon data collected by means of a third (post-election) wave of interviews. Funding was provided by a grant from the British Economic and Social Research Council.

4. The support of the Economic and Social Research Council (ESRC) is gratefully acknowledged. The work was funded by ESRC award number R000 23 2404.

5. Full details of the first wave survey are given in Rüdig et al. 1991.

6. The remainder of first wave respondents could not be identified for second wave mailing purposes, for example because the respondent had removed the relevant information from the first wave questionnaire before returning it.

7. Given the unwillingness of Green Party members in other European countries to participate in this type of survey which some see as 'un-green' (cf. Poguntke 1990; Kitschelt and Hellemans 1990), the response rates achieved are very satisfactory. In addition, we have to take account of the fact that the drop-out of members was already well-under way in late 1990 and disillusioned members leaving the Party were less likely to respond to a mailed questionnaire of this nature. This obviously had an adverse effect on response rates.
8. Respondents were invited to respond to a battery of possible reasons, indicating whether it was " Decisive; Very important but not decisive; Important; Not very important; or Played no role whatsoever" for their decision to leave the Party.
9. The exact question was: "The autumn conference at Wolverhampton in September 1991 approved a motion on the reform of the Party organisation, which includes the formation of an executive committee and the election of two principal speakers representing the Party to the outside. What is your view of the likely effect of these changes on the Party's future?" The aim was to enable respondents to give an opinion on the effects of these reforms independent of whether they had heard of Green 2000 or not.
10. Figures do not add up to 100% because of rounding.

References

Flynn and Lowe, P. (1992). 'The greening of the Tories: The Conservative Party and the environment', in Rüdig W. (ed.) *Green Politics Two* (Edinburgh: Edinburgh University Press), pp. 9-36.

Frankland, E.G. (1990). "Does Green Politics Have a Future in Britain? An American Perspective", in Rüdig W. (ed.), *Green Politics One* (Edinburgh: Edinburgh University Press), pp. 7-28.

Franklin, M.N. and Rüdig W. (1992a). "The green voter in the 1989 European elections", *Environmental Politics*, Vol. 1, No.4, pp. 129-159.

Franklin, M.N. and Rüdig W. (1992b). "On the durability of green voting: Evidence from the 1989 European elections". Unpublished paper.

Kitschelt, H. and Hellemans, S. (1990). *Beyond the European Left: Ideology and Political Action in the Belgian Ecology Parties* (Durham, N.C.: Duke University Press).

Mez, L. and Rüdig W. (1979). *Energiediskussion in Europa.* (Villingen: Neckar Verlag).

Poguntke, T. (1990). "Party activists versus voters: Are the German Greens losing touch with the electorate?", in Rüdig W. (ed.), *Green Politics One* (Edinburgh: Edinburgh University Press), pp. 29-46.

Reif, K. (1984). "National electoral cycles and European elections 1979 and 1984", *Electoral Studies,* Vol. 3, pp. 244-255.

Richardson, J.J.; Maloney, W.A. and Rüdig W. (1992). "The dynamics of

policy change: Lobbying and water privatisation", *Public Adminstration*, Vol. 70, pp. 157-175.
Rootes, C.A. (1991). "Environmentalism and Political Competition: the British Greens in the 1989 European Elections to the European Parliament", *Politics*, Vol.11, No.2, pp.39-44.
Rüdig, W. (1990). Anti-Nuclear Movements: A World Survey of Protest against Nuclear Energy (Harlow: Longman).
Rüdig, W. and Lowe, P. (1986)."The withered 'greening' of British politics: the case of the Ecology Party", *Political Studies*, Vol. 34, pp. 262-284.
Rüdig, W.; Bennie, L. and Franklin, M.N. (1991). *Green Party Members: A Profile* (Glasgow: Delta Publications).

3 Deep ecology and the undermining of green politics

John Barry

Introduction

Deep ecology has exerted a disproportionate influence on the development of green politics. Since Naess drew the distinction between 'The Shallow and the Deep, Long-Range Ecology Movement' (1973), green political theory and practice has been characterized by the type of splits and schisms one normally associates with the extreme left. From the infamous 'Fundi-Realo' division within the German Green party (Bahro, 1986, Capra and Spratnak, 1986), and other European green political parties, (Doherty, 1992), to the 'deep ecology' versus 'social ecology' debates within the U.S. green movement, (Chase (ed), 1991), the development of an integrated and united green politics and ideology has been problematic. At both the pragmatic and theoretical levels deep ecology seems to have been responsible for more harm than good, creating a major ideological schism within green political theory and hampering its political effectiveness.

The aim of this paper is to present a critical analysis of deep ecology and its relationship to green political theory in general and European green discourse in particular. As I hope to show problems associated with it are such that it should be discarded as the philosophical foundation of green politics. I argue against the notion that "There must be no doubt that Deep Ecology is indeed the Green Movement's philosophical basis" (Dobson, 1989, p. 41), I also wish to show that to be a non-shallow green in Naess's sense, one does not need to ascribe to deep ecology and its spiritual or mystical metaphysics. Radical green politics does not need a commitment to the principles and philosophy of deep ecology, indeed such a commitment is I argue a regressive development in the attempt to create a feasible green political praxis and a coherent green political theory. This is not to say however that it should be banished altogether. It should be seen as belonging to what might be described as green moral psychology, the personal and motivational 'springs of action' of some greens. It is thus a contingent rather than an essential element of green politics.

The starting point of this critique of deep ecology is Dobson's statement that "The politics of ecology does not follow the same ground rules as its philosophy." (1990, p. 68). So long as deep ecology in its various forms is held to be the philosophical grounding of green politics this will always be so. I hope to show that some of the problems associated with green politics stem from the undeserved emphasis often put on deep ecology, and that it creates a false and misleading dichotomy between 'deep' and 'shallow' greens. The danger in prioritizing deep ecology is that green theory runs the risk of sacrificing its democratic imperative for the purity of its eco-philosophical critique, and abandoning the positive potentials of modernity in the mistaken belief that it is irredeemably flawed. As a more positive contribution to the debate, an alternative philosophical and moral basis for green politics is outlined.

Overview of deep ecology

Much of the confusion concerning deep ecology and its problematic relationship with the wider green political project, has got to do with the fact that deep ecology was primarily an ethical theory in its early stages of development. The use of the adjective 'deep' by Naess was to distinguish his conception of the central concerns and motivations of the radical ecology movement from the 'shallow' movement, which Naess characterized (with hindsight, some would say caricatured) as concerned with the "Fight against pollution and resource depletion. Central objective : the health and affluence of people in the developed countries." (1973, p. 95). The 'deep' of deep ecology referred to the idea, still held by those who subscribe to its teachings, that the 'real' roots of the ecological crisis stems from the particular mind-set or paradigm with which we moderns view the world. As such deep ecologists are the modern representatives of a long line of counter-Enlightenment thought stretching back to Rousseau, Malthus, and the romantic poets (O'Riordan, 1981, pp. 23-60).

There are many problems with deep ecology, especially concerning the legitimacy of extending the self to include the non-human world (Mathews, 1991, Naess, 1989, Fox, 1990), and the idea of 'biocentric egalitarianism' (Fox, 1984). The logical dilemmas of confusing our relation to the non-human world and advocating an ecological identification with it, in some 'deep' sense as proposed by deep ecology, has been noted by Sylvan: "Since you and I are both one with the planet, and you thin and I fat, you are both fat and thin, old and young and so on.", (1985, p. 17). For him the cause of some of its more bizarre consequences has to do with its 'extreme holism' (p. 10). Brennan has criticized what he calls 'mystical ecology' for advocating a non-discrimatory fusion of self and world, such that self-realization becomes co-extensive with the self-realization of the rest of the world. According to him, "The resulting morality is no improvement on anthropocentric morality; the radical extension of the agent is not a genuine extension of moral considerability.", (1988, p. 144). One could present the argument that deep ecology is motivated more by strategic concerns about the means to get to an *a priori* given, namely some 'innate' intuitionistic basis for the moral considerability of non-human nature, rather than giving reasons as to why we ought to extend our moral awareness to it. It is this that lies at the heart of Fox's advocating of intuition and self-identification as the way forward for the resolution of our environmental crisis: "For transpersonal ecologists, given a deep enough understanding of the way things are, the response of being inclined to care for the unfolding of the world in all its aspects follows 'naturally'- not as a logical consequence but as a psychological consequence; as an expression of the spontaneous unfolding (development, maturing) of the self." (1990, p. 247).[1]

The various ways put forward by deep ecologists as to bring about this psychological experiential 'unfolding of the world', range from: (a) advocating a return to a pre-modern type of ecological sensibility, hence the frequent references to the ecological outlook and epistemology of pre-modern peoples one finds in deep ecology, (Fox, 1986, p. 76, Chawla, 1991, p. 261), (b) the creation of an explicitly post-modern eco-consciousness (Cheney, 1990), or (c) some eclectic fusion of the two, (Capra, 1983). What seems to link all of these is a commitment to the idea that experience of nature be both 'hands on', experiential, or sensuous, and that nature be conceived as wilderness, as the 'Other'. For example Sale is clearly working with this view when he states that we must become 'dwellers in the land' "We must somehow live as close to [the land] as possible, be in touch with its particular soils, its waters, its winds; we must learn its ways...we must make its rhythms our patterns, its laws our guide", (1984, p. 224-5).[2] On this reading deep ecology can be considered as part of 'New Age' mysticism which attracts many greens, while it can also be used by critics as evidence that the real aim of greens is to return to some 'golden ecological age', turning the clock back to a time before the 'second fall' of humanity, as many deep ecologists characterize the Industrial Revolution.

Deep ecology and the re-enchantment of the world

This 're-enchantment' of the world is what I take to be the defining feature of contemporary deep ecology. Only by viewing it as having this thinly veiled agenda of generating a new cosmology, a re-spiritualization and/or 'rediscovery' of the mythic importance of nature, on an explicitly non- or at times anti-rationalist (and anti-humanist) basis, can we understand what deep ecology is really about, and how it leaves green politics bereft of anything substantially political. By making the green critique of the problems of modernity co-extensive with the disenchantment of the world as a result of the enlightenment, deep ecology makes green politics seem to be part of what has been called the 'tender hearted' critique of modernity. Deep ecology implies that green discourse is concerned with recovering the certainties of the past, when the non-human world was understood in terms of a cosmology that gave it (and life in general) meaning, value and significance. It therefore gives green politics a utopianism and an idealism that is neither necessary nor desirable.

The 're-enchantment' thesis of deep ecology comes from its particular understanding of the roots of the ecological crisis. The essence of the deep ecology position is that if the cause is a particularly environmentally unfriendly from of consciousness, then the logical thing to do is to advocate a more environmentally friendly form. Within the deep ecology literature this usually takes the form of a spiritual, religious or mystical attitude towards the world. Lynn White set the tone, when he conceived of the ecological crisis in terms of the Christian view of the world. For him it seemed axiomatic that "Since the roots of our troubles are so largely religious, the remedy must also be essentially religious, whether we call it that or not.", (Barbour (ed), 1983, p. 30). We can say that deep ecology represents a similar 'paradigm shift', in that it advocates abandoning conventional moral discourse and urges the search for an alternative paradigm, epistemology, and cosmology. In short a completely new *weltanschauung* is proposed to replace the Western, rationalist one. Fox sums up this idea, evidencing the elevation of intuition and the lack of a political strategy that characterizes deep ecology, when he states: "Deep ecologists are willing to trust their inner voices in the hope that the dominant social paradigm (within which the moral community is situated) will disintegrate." (1984, p. 196).

The dangers of this are extremely serious for green politics. By citing reason and rationality as the culprits of the ecological crisis, by emphasizing how the process of rationalization disenchanted the world and thus left it open to be considered purely instrumentally, deep ecology can be argued to risk throwing the baby out with the bathwater. By criticizing reason per se, rather than locating the real cause in a hyperactive and colonizing instrumental rationality, as Habermas does (1987); or in the ecologically destructive effects of economic rationality, as Gorz does (1989), deep ecology denigrates reason and modernity, and confuses the crucial distinction between the concept and a particular conception of reason. By emphasizing the importance of intuition, myth, and what Fox calls a more "enchanting cosmology than mechanistic materialism." (1984, p. 195), it denies green politics any intersubjective critical standards by which we can judge human interaction with our environment. Mystical or spiritual orientations are beyond verification, they require faith, not reason, as Fox's statement above indicates. In other words there are no agreed standards against which one can judge or rank deep ecological 'enchanted' orientations to 'being in the world'. This is because there is an implicit egalitarianism between cosmologies, one is as good as another in a non-consequentialist sense. Deep ecology is non-consequentialist in part because it is also non-anthropocentric. In general, it likes to think of itself as biocentric (Naess, 1973, 1989) or ecocentric (Fox, 1990, Eckersley, 1992), and for it reasons for care are as important as the care itself (Dobson, 1989, p 46).

The real question surrounding this re-enchantment thesis of deep ecology is, not as Dobson seems to think, that it offers a way for greens to "[shift] the onus of justification from those who would preserve the non-human world to those who would intervene in it", (1990, pp. 64-5), but the more basic issue of how to argue this position and convince people to accept it as part of a new normative orientation to the world. For example, which eco-cosmology and moral psychology are greens to accept: the view put forward by Capra with its blending of quantum physics and eastern mysticism (1983); Smith's recovery of a Heideggerian sense of 'Dasein' or 'being' (1989, p. 6); Fox's view that the ecological crisis will not be resolved until "we all begin to think more or less like Smohalla [a Native American Indian tribe].", (in Dobson, 1989, p. 44); Naess's 'Ecosophy T' with its overarching principle of 'self-realization', (1989, p. 196-212), or Mathews' 'ecological sense of self' (1992)? In the last analysis there is nothing in deep ecology to help us judge between these different ways of 'being in the world'.

The thrust of this critique of the 're-enchantment' strategy both as a way of acknowledging the moral significance of the non-human world, and as an adequate response to finding ways to create a more harmonious relationship with it, is that it is on both counts undesirable and untenable. One of its most undesirable features is this: who is to 're-enchant' our meaningless environment? Deep ecology mystics who have seen the (green) light ? Are they to become a new benign elite in the mould of Plato's guardians looking after the interests of their human and non-human flocks within what Plato called the 'healthy state', or what deep ecologists today would perhaps refer to as the 'bioregional state', (Sale, 1980, 1984, pp. 224-50) ? We can see that the deep ecology answer to our eco-crisis is anti-Enlightenment in another damaging sense for green politics: it has the potential for the creation of a profoundly anti-democratic politics (Bramwell, 1989, chapter. 8, Dryzek, 1990, p. 200).[3] For a politics that holds democracy and participation to be its *sine qua non*, having such anti-democratic potentials at its foundations is unacceptable. Unlike the attempt to include non-humans in our moral discourse (i.e., anthropocentrically), mystical or cosmological ontologies seem to generate a particular spiritualized conception of green politics, such that to be 'really' green one must subscribe to some conative or intuitive epistemological understanding of the non-human world and our relationship to it. An example of this is Porritt's conclusion to his highly popular introduction to green politics; "To the enduring challenge of social justice we *must* now add the challenge of spiritual enlightenment", (1984, p. 233, emphasis added). Why 'must' ? Is part of the green project to 'force people to be spiritual' ? The problem for green politics is that the commitment to 'spiritual enlightenment' could undermine its concern for 'social justice'.

In an age of secularization on the one hand and the growth of religious fundamentalism an the other, deep ecology and its various cosmologies present green politics as just another competing variant of 'millenarianism': "The belief and practices of those who seek by way of a religious and/or political movement, to secure a comprehensive, salvationary solution for social, personal, and political predicaments.", (Bullock and Stallybrass (eds),

1978, p. 390). In fact deep ecology is by and large apolitical, viewing the issue of how to achieve the sustainable society as something residual, parasitically dependent on specific forms of consciousness. This leads to a 'greener than thou' moralistic tone, typified by Porritt's statement that: "Some kind of spiritual commitment or religion in its true meaning (namely the reconnection between each of us and the source of all life), is a fundamental part of the transformation that ecologists are talking about.", (*op. cit.*, p. 211), or Dobson, discussing Bahro (1986), to explain that; "The general point about the religious approach is that the changes that need to take place are too profound to be dealt with in the political arena." (1990, p. 143). If both of these leading commentators of modern European green political theory are right, I think green politics is in grave trouble, and runs the very real risk of becoming just another Quixotic element of the 'New Age' caravan; a salvationary cosmology combined with a doomsaying moralism that undermine its political ambitions. This re-enchantment strategy lapses into the contentious area of definitions of the 'good life', and leaves green theory open to criticism that it is both backward looking and perfectionist. With deep ecology, green politics, contrary to its professed statements, has both anti-democratic and illiberal potentials.

Deep ecology: Green politics without the politics

Perhaps the greatest disservice deep ecology does to green politics is that it denies the crucial importance of political strategy and the need to confront the state and vested interests in the transition to a sustainable society. Because it sees consciousness as the both the problem and the solution, green politics becomes at best a stepping stone for its re-enchantment project or, at worst, a residual sphere, in the sense implied by Bahro where green politics really is deeper than politics. As Bookchin puts it, "New Age mystics and deep ecology acolytes do not 'confront', be it the corporate establishment, bourgeois social relations or the omnipresent state.", (in Elkins, 1989, p. 49). What is needed is not the wholesale rejection but a recognition that a precondition for a coherent and realistic green politics is a self-limiting deep ecology i.e., its explicit confinement to the realm of green moral psychology. In its present form, deep ecology is neither concerned with ethics per se, nor with politics, but with metaphysics, and New Age 'personal development'. It should therefore be considered a contingent, optional element of political ecology, one with no political or policy implications.

That deep ecology is politically naive when it comes to discussions of the reality of political conflict can be readily seen by the eclectic mix of strategies advocated by Naess (1989). His advice for greens is that "There has to be an intimate cooperation between conservationists, journalists, people knowing political ways and means, and all those who are vitally interested in the workings of big societies", (1989, p. 131). Hardly the type of political strategy one would think a radical politics would endorse! In a similar vein Naess states in a most generalized form "I envisage a change of revolutionary depth and size by means of many smaller steps in a radically new direction... The direction is revolutionary, the steps are reformatory.", (*ibid*. p. 156. As to what either of these statements of 'strategy' mean or entail is never spelt out, thus leaving us none the wiser as to his political tactics. According to Dryzek, this should not surprize us, "Most of those who speak of eco-spirituality say little about this transition, let alone any practical political programme", (1989, p. 201). Thus we can criticize it for not emphasizing and prioritizing the political, either in the search for a green historical agent, (Dobson, 1990, pp. 156-70), the importance of the capitalist state in any discussion of green politics, (Carter, 1990, p. 4, O'Connor, 1990, p. 2), or the basic Marxian idea that 'social being' (partly) determines consciousness, (Dobson, 1989, p. 46). Without this kind of theoretical work political ecology may be undermined by the prioritizing of the deep ecology agenda over that of green politics.

By focusing on the effects of our ecological crisis, our destruction of the biosphere, deep ecology has relatively little to say about its distinctly political and economic underlying causes (Barry, 1990). Implicit in the attention given to the purely ecological dimension of the 'crisis of modernity', is the tension between human 'social justice', especially in the radical democratization greens see as part of the process of achieving this (Carter, 1991, pp. 12-15); and accepting the view that 'nature knows best', which commits deep ecology to saying that we must 'read off' social practices from nature. That a full-bloodied acceptance of the latter may undermine the former is clearly acknowledged by Sale when he accepts the 'nature knows best' position, stating that: "Political principles on a bioregional scale are also grounded in the dictates of nature... It does not mean that every region ... will construct itself upon the values of democracy, equality, liberty, freedom, justice, and other such like desiderata", (1984, p. 312). It seems that prioritization of the ecological (including eco-consciousness raising) over the social or the political subverts green claims to be a radical democratic ideology. If we are to live according to the dictates of nature, as Saward (1991) and Ferris (1991) point out, and the quote from Sale shows, there is a strong possibility that democracy becomes an optional extra. The point here is that the undermining of a green conception of the political, and its democratic ambitions is perpetrated by an obligation to the esoteric agenda of deep ecology in what might be called the subversion of green politics by metaphysics.

Nature as wilderness and garden [4]

The 'wilderness experiences' advocated by many deep ecologists like Devall and Sessions (1985), and eco-feminists like Warren (1990) are arguably an aesthetic form of 'green consumerism' (to which deep ecologists and most greens are formally opposed) as the way to create a new eco-friendly consciousness. Both are individualistic, they appeal to the person as consumer or experiencing subject, not as a citizen, thus both are apolitical, and largely a function of income distribution. This is especially true in the case of 'wilderness experiences'. It also gives another elitist dimension to deep ecology exacerbating the one mentioned above. The elitism here comes from the fact that deep ecologists do not think that the experience of nature one gets from working on an allotment, for example, is of the same consciousness raising quality as, say, Warren's personal experience of climbing a rockface (*ibid.*, pp. 134-5). The obvious bias here is that only those who can afford these experiences will truly have their consciousness raised, leaving the rest of us perhaps to have a 'mediated' eco-consciousness authoritatively inculcated as indicated above. On the other hand perhaps Elkins is nearer the mark, and offers a less 'draconian' interpretation, when he asserts that for deep ecologists: "Since social and ecological problems are ultimately seen to be the consequences of misconceived values and ideas, social change is contingent on the ability to communicate the insights of the new paradigm." (1989, p. 62). But it seems that this more benign interpretation is purchased at the cost of rendering deep ecology politically impotent, leaving it as a sort of whispering campaign against the modern worldview.

A more obvious practical problem concerning 'wilderness experience' as a strategy even for consciousness-raising purposes, relates to people's actual access to wilderness. Access to wild nature is increasingly what Hirsch (1976) has called a 'positional good', where the zero-sum nature of the good means that only those with progressively increasing amounts of income can enjoy them. The whole point about positionality is that the more people who have access to a positional good the more that good loses its value. Thus encouraging modern urban populations en mass to 'contemplate a mountain' will lead to the diminution of the 'consciousness-raising' potential of the experience. Thus even when deep ecologists do talk about 'strategy', (although there is nothing political about isolated 'wilderness experiences') it is so fraught with inconsistencies and naïvete, that it is better abandoned as an integral part of green political theory and practice.

This access to wilderness also has to do with its distribution across the globe. For example, it is clear that in the western European context there is little 'wilderness' in the requisite sense required by deep ecologists. In Europe a conception of nature as 'garden' seems more appropriate, given the fact that humans have largely made the modern European landscape, from the Dutch reclaiming land from the sea, to the hedgerows of the English and Irish countryside. In a very real sense then there not much 'wilderness' left in Europe. Therefore to talk of 'wilderness experience' is to talk in a foreign language when addressing the populations of Europe. There can be no going back to a pre-humanized environment, and to suggest attempting to do so resonates with nostalgia and utopianism.

Outline of an alternative to deep ecology

I hope to have shown that the problems associated with deep ecology are simply too great and that its advantages in motivating certain people to care about the environment are more than outweighed by the various difficulties it causes when seen as the philosophical or moral basis of green politics. I do not accept the idea that moral consideration of the environment requires either a return to some 'enchanted' pre-modern cosmology, or a 'New Age', post-modern ecological sensibility. One can accept much of their critique of instrumental rationality, but reject their aim of criticizing the entire Enlightenment legacy.

There is ample scope within conventional moral discourse for us to include certain parts of the non-human world in the moral community. The starting point for a realistic green moral theory must be a rejection of equal intrinsic value that deep ecology implies. For example, I do not consider it an instance of the 'arrogance of humanism' (Ehrenfeld, 1980) to say that if it is a choice between the life of a human and the destruction of a tree, or a cow, there is a *prima facie* presumption in favour of the human, such that if the latter is sacrificed for the former, we would rightly call it morally perverse and wrong. In regard to future human generations, this is a real rather than a hypothetical issue, and one that should be a central element in a coherent green theory, i.e., one with a better 'fit' between ethical consideration of the non-human world, and the emancipatory political goals of green politics.[5] What is perhaps required is the extension of moral considerability to nature, while maintaining that full moral agency is the sole privilege of presently existing humans. In this way non-humanity is not totally devoid of moral worth, but the pitfalls and counter-intuitive results of deep ecology are avoided. This 'ethic of use' suggests certain moral limitations to our interaction with the environment, but it does not necessitate the development of a contentious 'verdant spirituality' or 'ecological sense of self'. One of the biggest problems with deep ecology is that it conceived the ascribing of value in a sort of zero-sum fashion. Acknowledging human beings as the only valuing subjects on earth, and ourselves as being more valuable than non-humans, meant the complete denial of moral considerability to the latter. This precluded the possibility of there being a moral space between a crude instrumental view of nature, and a completely non-instrumental mysticism.

Another issue that needs to be considered in the formulation of an alternative to deep ecology, is the fact that the starting point of any environmental ethics should be the very real issue of moral deliberation and right conduct when the interests of ourselves and non-humans conflict, rather than the meta-ethical considerations that so consumed the early development of green moral theory. Here, I think we need to bring in concern for future generations, as encapsulated in the idea of 'sustainability'; the vexed issues of human basic versus non-basic needs or interests and, the interests of non-humans, such that we can say the basic interests of humans over-ride those of non-humans.[6] Realistically we must recognize that only certain parts of non-human nature will be accorded moral worth, and that we should not worry unduly about this. As Fox (1984) noted, cows do scream louder than carrots, and presumably carrots louder than amoebas. There is thus an ineliminable anthropocentrism at the heart of this alternative environmental ethic, that does not imply a categorical denial of moral considerability or non-instrumental value to certain parts of nature. Dobson seems to accept this 'ineliminable anthropocentrism' when he draws a distinction between 'weak' and 'strong' anthropocentrism (1990, pp. 63-71), holding that "there is room for a (weak) anthropocentrism in respectable ecological statements. The reason for this is that weak anthropocentrism is a necessary feature of the human condition", (p. 65). Indeed, we can go further and posit that this 'weak anthropocentrism' is an urgently required imperative to be integrated within green political discourse. This 'weak anthropocentrism' accepts that we may be the only valuing entities but cautions against the arrogant presumption that we are the only morally valuable entities, on the one hand, and avoids the problematic assertion that humans and non-humans are of equal moral value, on the other. It also steers clear of the temptation to simply erase the boundaries between ourselves and non-humanity as a way of asserting the moral status of the latter. It thus respects the 'otherness' of nature.

The important areas of development for a non-deep ecological green moral philosophy, concern finding more 'humane' ways to fulfil our basic needs, and our obligations to future generations. One of the advantages of this starting point is that there is a correlation between concern for future generations and concern for non-humans. Anthropocentric reasons for changing our practices with regard to the non-human world may restrict our destructive interference with it, and would lead to the interests of non-humans being protected. From a consequentialist and non-human perspective, cows, we may conjecture, are not too concerned as to the reasons we give for not eating them: what matters to them is that we do not actually eat them. Although the reasons may be anthropocentric, the results do not have to be.[7] If green politics and morality are to have some sort of theoretical coherence and strategic import, it is better that we construct a moral theory that is in harmony with the politics, and not, like deep ecology undermine it for the sake of philosophical purity. In the dialogue between it and green politics, the former is obliged to accommodate itself to the latter, not vice versa. It behoves greens who are seriously concerned about political change, to recognize what Carter terms the "primacy of the political" (1990, p. 8). Together with the non-instrumental anthropocentrism outlined above, this conception of green politics stands a better chance of achieving the harmony and non-destructive interference with the non-human world that deep ecology seeks.

Following Habermas, Dryzek (1990, 1992) and Gorz (1989), I would argue that the ecological crisis is not primarily as a result of rationality *per se*, but due to the one sided development of rationality, such that technical-instrumental rationality has become almost ubiquitous and has left the non-human world barren of any ethical significance. The price paid for our progressive liberation from material want, poverty, and superstition has been the virtual eradication of our interaction with nature as a (partly) moral relationship, and it becoming an almost purely technical matter. [8] But to rectify this we do not need to rush into the first or most attractive eco-cosmology or a particular hue of Dobson's 'verdant spirituality' (Dobson (ed), 1991, p. 9). Accepting that the denial of any ethical significance was instrumental in creating the present crisis, does not mean that embracing an expanded conception of the self, for example, will bring about the ecological status quo ante that is supposed to have held prior to the Enlightenment. Neither does it mean that reason and rationality need be jettisoned in favour of what Dryzek calls "a regressive emphasis on spirit.", (1990, p. 199). In fact the ecological crisis is argued by Gorz, for example, as being due to the irrationality of modern economic and social systems, "The current crisis is not the crisis of Reason but that of the (increasingly apparent) irrational motives of rationalization as it has been pursued thus far.", (1989, p. 1). What is needed therefore, is not a return to some pre-modern ecological sensibility, or the creation of some post-modern cosmology, but the confining of economic rationality, and other instances of instrumental rationality to their proper sphere, by arguing for the primacy of what Dryzek terms "The ecological rationality of discursive politics", (1992, pp. 38-42), which allows an ethical significance to nature without lapsing into an irrational mysticism.

Conclusion: The legacy of modernity

As far as the critique of the 'dominant social paradigm' goes, I do not accept that the understanding of the non-human world yielded by science leaves us with a totally disenchanted world, nor does it somehow lead us automatically to an arrogant humanism that sees our species as somehow 'above' or 'beyond' nature. I would tend to agree with Grey in his critique of deep ecology, that "A purely secular, scientific naturalism can provide a thoroughly satisfying way of realizing our unity with the non-human world", (1986, p. 212). There is nothing essentially anti-natural in science that leads to a radical break in our link to the natural world. This is another injury deep ecology does to green politics: caricaturing it as being some sort of anti-science ideology, usually together with a Luddite view of technology. Once again this gives critics an unnecessary opportunity to reject it, and increasing the probability that others more sympathetic will misperceive its basic aspirations.

In terms of epistemology, the type of knowledge generated by a scientific understanding of the world can bring home to us just how dependent we are on it. Scientific knowledge enhances our relationship to our world. We are in more intimate, closer contact with nature when we understand more not less. It may well turn out that the greatest achievement of the Enlightenment and the scientific world view that underpinned it, is not that it gave us certain types of knowledge and technologies to control nature, but that for us in the last decade of the twentieth century, it acknowledges our complete dependence on it, our pressing need to find some harmony with it, and above all shows that what unites us with nature is prior to what separates us.

From the point of view of green theory, it is within the Enlightenment-derived democratic tradition that it places itself in its arguments for participatory democracy and decentralization of the state and civil society. Deep ecology is an anti-modernist philosophy, seeking to re-enchant the world and return to some pre-modern ecologically sensitive cosmology. It also depends on a particular conception of 'nature as wilderness' for its consciousness-raising project. Whatever its merits in countries and regions where it was formulated, where 'wilderness' is a real as opposed to an abstract concept, the conclusion we must reach is that it has little relevance to those for whom nature as 'garden' is their lived reality. If we accept the 'primacy of the political' this implies that green political theory emphasize its democratic agenda, and seek to accommodate green philosophy to that imperative. What the analysis seems to suggest is that deep ecology is not relevant to the development of the green political project, indeed, as I hope to have shown, it represents a regressive move. The project of modernity is not irredeemably flawed, as deep ecology implies. Greens need to realize that it offers resources necessary to create a sustainable and just society: in other words it holds out the possibility of reforming itself, and realizing its emancipatory promises. In rejecting deep ecology, green politics should consider itself as part of this self-critical, self-reforming Enlightenment tradition.

Footnotes

[1]. This emphasis on 'maturing of the self' obviously implies that Fox has some conception of the self and its 'proper' development. It seems that from the deep ecology point of view those that do not adopt its position are somehow 'immature' or 'undeveloped'. This ethical 'naturalism' resonates throughout green theory in general, and deep ecology in particular.

[2]. In this respect deep ecologists literally adhere to Commoner's 'third law of ecology', namely 'nature knows best' (1971, p. 23). Although Commoner coined this phrase to indicate the complexity of the ecosystem, and our inability to completely understand it, and predict the full consequences of our interference with it; deep ecologists go further and argue that we ought to 'read off' intra-human practices from nature. What deep ecology implies is that a more 'natural' way of life is not just more ecologically sensible, but is a better way of life for humans.

[3]. Criticism of the direct action group Earth First!, according to Dobson "the political wing of deep ecology" (1990, p. 102), have been centred on its authoritarianism and anti-humanism. See Bookchin's critique in Chase (ed) (1991).

[4]. For the most part deep ecology writers come from nations and regions where 'wilderness' is a reality rather than an abstract concept; Australia (Fox, Eckersley), west coast America (Sale, Earth First !), or Scandinavia (Naess). For an analysis of the distinction between nature as 'wilderness' and 'garden' see Nash (1981, pp. 81-91). Although outside the scope of this paper, there are arguments that can be raised

about the whole idea of nature as wilderness. One of the most compelling is that human-induced global ecological problems have effectively meant that there is no where on the planet untouched by humanity. In Marxian terminology, we have 'humanized nature', although in a negative rather than positive way.

[5]. Because deep ecologists are more concerned with non-humans, they have relatively little to say on the issue of future human generations, or indeed global distributive justice across the presently existing global population.

[6]. This raises the issue that often crops up within green politics, between the idea of the sustainable society, and the principle of sustainability. The moot point here is that the principle merits more attention than describing and prescribing ecological utopias (see Barry, 1993).

[7]. Brennan cautions against thinking that extending moral concern to non-humanity will obviate our instrumental use of it."Even if morality succeeds as a device for counteracting limited sympathies within the human community, it is unlikely to succeed as a device that will enable us to yield priority over human concerns and interests to the good of things 'natural, wild and free'.", (1988, p. 30).

[8]. The problem with Habermas's theory is that he seems to allow for no moral dimension to our interaction with the environment. The whole point of his 'discourse ethics' is to prevent our relations with each other becoming like those with nature. He goes so far in his anthropocentrism to deny any ontological distinction between living and inanimate nature, (Keane, 1985. p. 334). For a green critique of Habermas see Eckersley (1992, chapter. 4).

Bibliography and references

Bahro, R. 1986. Building the Green Movement, (London: Heretic Books).
Barry, J. 1990. 'Red and Green', in *Making Sense: Ireland's Political and Cultural Review,* Summer, pp. 10-15.
Barry, J. 1993. *'Towards a theory of the green state'*, (forthcoming).
Bookchin, M. 1986. *The Modern Crisis,* (Philadelphia: New Society).
Bramwell, A. 1989. Ecology in the 20th Century: A History, (Cambridge: Cambridge University Press).
Brennan, A. 1988. Thinking about Nature: an Investigation of Nature, *Value and Ecology,* (London: Routledge)
Bullock, S and Stallybrass, F. (eds). 1978. *The Fontana Dictionary of Modern Thought,* (London: Fontana).
Capra, F. 1983. *The Turning Point: Science, Society and the Rising Culture*, (London: Fontana).

Capra, F and Spratnak, C. 1983. *Green Politics: The Global Promise,* (London: Pion).
Carter, A. *'Toward a Green Political Theory',* (Forthcoming).
Chase, S. (ed). 1991. *Defending the Earth: A Dialogue between Murray Bookchin and Dave Foreman,* (Boston: South End Press).
Chawla, S. 1991. 'Linguistic and philosophical roots of our environmental crisis', *Environmental Ethics,* 13:3, pp. 253-62.
Cheney, J. 1989. 'Post-modern environmental ethics- ethics as a bioregional narrative', *Environmental Ethics,* 11:2, pp. 117-35.
Commoner, B. 1971. *The Closing Circle: Nature, Man and Technology*, (New York: Bantam).
Devall, B and Sessions G. 1985. *Deep Ecology: Living as if Nature Mattered,* (Layton, Utah: Gibbs M. Smith)
Dobson, A. 1989. 'Deep Ecology', in *Cogito*, (Spring), pp.41-6.
Dobson, A. 1990. *Green Political Thought*, (London: Unwin Hyman).
Dobson, A. (ed). 1991. *The Green Reader*, (London: Andre Deutsch).
Dobson, A. 1992. 'Review article: Nature in question', *Environmental Politics,* 1:1, pp. 144-8.
Doherty, B. 1992. 'The Fundi-Realo controversy : an analysis of four European Green parties', in *Environmental Politics,* 1:1, pp. 95-120.
Dryzek, J. 1990. 'Green reason: communicative ethics for the biosphere', in *Environmental Ethics*, 12:3, pp. 195-210.
Dryzek, J. 1992. 'Ecology and discursive democracy: Beyond liberal capitalism and the administrative state', in *Capitalism, Nature, Socialism,* 3:2, pp. 18-42.
Eckersley, R. 1992. *Environmentalism and Political Theory: Toward an Ecocentric Approach,* (London: University London College Press).
Ehrenfeld, D. 1981. *The Arrogance of Humanism,* (Oxford: Oxford University Press).
Ferris, J. 1991. *'Ecological versus social rationality: Can there be green social policies',* (Forthcoming).
Fox, W. 1984. 'Deep Ecology: A new philosophy of our time ?', in *The Ecologist,* 14:4/5, pp. 194-200.
Fox, W. 1990. *Toward a Transpersonal Ecology: Developing New Foundations for Environmentalism,* (Boston: Shambhala).
Goodin, R. 1992. 'The high ground is Green', in *Environmental Politics,* 1:1, pp. 1-9.
Grey, W. 1986. 'A critique of deep ecology', in *Journal of Applied Philosophy*, 3:2, pp. 211-6.
Gorz, A. 1989. *Critique of Economic Reason,* (London: Verso).
Habermas, J. 1987. *The Theory of Communicative Action.* Vol. 2., (Boston: Beacon).
Hayward, T. 1990. 'Ecosocialism- Utopian and Scientific', in *Radical Philosophy*, 56, pp. 2-15.
Hirsch, F. 1976. *Social Limits to Growth,* (Cambridge: Harvard University Press).

Luke, T. 1988. 'The dreams of deep ecology', in *Telos,* 76, pp. 65-93.
Mathews, F. 1992. *The Ecological Self,* (London: Routledge).
Naess, A. 1973. 'The shallow and the deep, long-range ecology movements', in *Inquiry,* 16, pp. 95-100.
Naess, A. 1987. 'Self-realization: An ecological approach to Being in the world', *The Trumpeter,* 4, pp. 35-42.
Naess, A. 1989. *Ecology, Community and Lifestyle,* (Cambridge: Cambridge University Press).
Nash, R. 1981. *Wilderness and the American Mind,* (New Haven: Yale University Press).
O'Connor, J. 1990. Conference Papers, (Santa Cruz: CES/CNS Pamphlet No.1).
O'Riordan, T. 1981. *Environmentalism,* (London: Pion).
Pepper, D. 1984. *The Roots of Modern Environmentalism,* (London: Croom Helm).
Porritt, J. 1986. *Seeing Green,* (Oxford: Basil Blackwell).
Sale, K. 1984. 'Mother of all: an introduction to bioregionalism', in Kumar, K. (ed), *The Schumacher Lectures: Vol. 2,* (London: Blond and Briggs).
Saward, M. 1991. *'Green Democracy?',* (Forthcoming).
Smith, N. 1989. 'On deep ecology', in *Argument,* 1:1, pp. 1-8.
Spratnak, C. 1986. *The Spiritual Dimension of Green Politics,* (Santa Fe: Bear & Co).
Sylvan, R. 1985. 'A critique of deep ecology: Part 2', in *Radical Philosophy,* 41, pp. 10-23.
Warren, K. 1990. 'The power and the promise of ecological feminism', in *Environmental Ethics*, 12:2, pp. 125-46.
White, Lynn, Jr. (1967). 'The historical roots of our ecologic crisis', reprinted in Barbour, I, (ed). 1984. *Western Man and Environmental Ethics,* (London: Fontana).
Zimmerman, 1987. 'Feminism, deep ecology and environmental ethics', in *Environmental Ethics*, 9:1, pp. 21-44.

Part II

BUSINESS, FINANCE AND ENVIRONMENTAL REGULATION

4 Environmental Assessment - The inadequacies of law

John Alder

Introduction*

In 1988, after many years of debate and after it had been through twenty drafts, the United Kingdom implemented the European Economic Community's Directive on Environmental Impact Assessment.[1] The Directive required members states to put into place by July 3 1988, procedures for assessing the environmental effects of certain land use projects. The Directive permits member states to incorporate environmental impact assessment into their existing procedures for development consent. The United Kingdom has chosen to do this, rather than to set up new machinery. During the protracted debates on the Directive the UK had argued that specific procedures for environmental assessment were both unnecessary and undesirable. They were said to be unnecessary because the UK planning law was sufficiently flexible and already required decision makers to take environmental matters into account and undesirable because of the risk of delay and expense.[2] There is also a strong tradition in English administrative culture of relying on discretion rather than law.

The purpose of this paper is to suggest that the general framework of English Law is inadequate to secure the aims of the Directive and that English legal culture is hostile to regulation of this kind, and indeed unsympathetic to environmental values. The handful of cases brought by environmentalists against what they considered to be inadequate implementation of the Directive have all been unsuccessful. In particular the courts appear to apply traditional English judicial review principles without taking account of the broader obligation to interpret legislation implementing European policies in the light of those policies.

The 1985 Directive was part of the 3rd Action Programme of the European Community. The main principles of this were "prevention is better than cure" and "the polluter pays".[3] English law, with its traditional emphasis upon

* *This chapter appears in The Journal of Environmental Law (1993) vol 5 no 2.*

remedies after the event is not well equipped to deal with the first principle. On the other hand, lacking a concept of constitutionally protected property rights English law should have less difficulty with the second. There is however a vein of judicial authority that 'public costs' should not be borne by private developers unless there is a *quid pro quo*.[4]

This elevation of the private over the public is reflected in the approach of both common law and equity to environmental issues. For example, unless there is "sensible injury to the value of property" nuisance law is aimed merely at preserving the status quo and lacks an objective method of allocating costs and benefits between private and public interests.[5] Similarly restrictive covenants as proprietary interests in land can restrain environmental damage but cannot require positive steps to be taken to improve the environment including clean up measures.[6] Moreover a covenant can be enforced only by a person whose proprietary rights are affected by the breach.[7]

The Environmental Impact Assessment Directive was introduced in 1985. This was before the Single European Act of 1986 altered the Treaty of Rome so as to make environmental protection a Community objective in its own right.[8] The 1985 Directive was enacted under Article 100 of the Treaty of Rome in order to remove unfavourable competitive conditions[9] and also, more dubiously, under Article 235 which provides for the fulfilment of community objectives - in this case "the protection of the environment and the quality of life" in cases where the Treaty does not provide specific machinery.

The Directive was implemented in English law by a series of statutory instruments made under the European Communities Act 1972 section 2(2).[10] This requires a strict transfer of the provision of the directive without any additional material and has to be interpreted in the light of the Directive. There is therefore the possibility of a clash between the traditional narrow semantic approach of English lawyers to questions of legislative construction and the broader, purposive approach practised in continental Europe and by the European Court. However the European Communities Act 1972 sections 2(4) and 3(1) requires our courts to accept as authoritative the decisions and principles of the European Court of Justice in relation both to the meaning and application of European Law. When interpreting community law itself or domestic instruments implementing community law, the courts must therefore reason in the manner of the European Court, that is to say they must take account of the social and political context and have regard to the general spirit purposes and concepts of European Community Law. The different language texts can be compared and the history of the Directive can be looked at as an aid to interpretation as can European environmental policy generally.[11] Much of this is unfamiliar if not uncongenial to English lawyers being even wider than the purposive approach of English law which involves a search for historic legislative intention. Nevertheless the European Court has recently made it clear that domestic law must be interpreted to conform to EC requirements[12] and Article 5 of the Treaty of Rome imposes a general obligation upon member states to give effect to the objects of the community. This is directly binding on U.K. courts by virtue of section 3 of the European Communities Act 1972.

Many of the regulations implementing the Directive concern projects carried out by public bodies, a path pioneered in the USA.[13] The Directive breaks new ground in that it also applies to projects carried out by private developers. This paper will concentrate upon private sector projects because it is here that the problems of integration into the general law are most acute.

Characteristically of European Community legislation the Directive is drafted in broad and vague terms. The Directive is particularly vague on matters of public participation and also upon the initial issues of scoping and screening which environmentalists regard as central to the enterprise of environmental impact assessment. The Directive fails to define the term environment at all and the central concept of 'project' is only loosely defined (Article 1). The Directive also gives member states considerable discretion as to matters of implementation. The Directive itself may therefore partly be to blame for the inconsistencies of application. Competition policy requires uniformity of application throughout the member states so as to create a level playing field. The Directive does not achieve this. Indeed it allows member states to impose more stringent environmental assessment rules.[14]

In relation to private sector projects English law has implemented the Directive by incorporating its requirements into the existing procedures for applying for planning permission for the development of land. Local planning authorities are required to take into account in deciding whether to grant planning permission, 'environmental information' comprising a written "environmental statement" provided by the developer and information provided by other public bodies and the general public. There is no requirement for a formal 'assessment' as such, but without an environmental statement planning permission cannot be granted.[15]

The requirements of the directive

The notion of an environmental assessment is nebulous. An assessment consists of a procedure prescribed in outline by the Directive under which certain environmental information must be taken into account before approval is given by national authorities to the implementation of those 'projects' which fall within the directive. The assessment can be carried out by the same authority that decides whether or not to approve the project. It need not be in writing.

An 'assessment' consists of three stages. Firstly the developer must at its own expense provide written information. This is prescribed in general terms by the Directive. It must 'at least' include the nature of the project, the measures envisaged in order to "reduce and if possible remedy" 'significant' environmental effects, data required to identify and assess the 'main' likely environmental effects and, very importantly, a 'non technical summary' (Article 5.2). A more detailed catalogue of information is set out in Annex 3.

This includes for example, alternatives, forecasting methods, technical difficulties and long term, indirect and secondary effects of the project. The Directive is unclear as to how compulsory this further information is. It certainly allows member states some discretion in deciding whether it is appropriate or reasonable to require this further information (Article 5.1). As we shall see the U.K. has taken advantage of this uncertainty. There are no scoping requirements, apart from a general list of the broad subject matter (Article 3)[16], Public bodies must on request make information available to the developer for the purpose of preparing the environmental statement but the developer is not obliged to consult anyone[17] It is left to the responsible authority to decide whether the Environmental Statement is sufficiently stringent. The obvious dangers of bias, failure to consult before submission, and inadequate expertise within local authorities, appear to have materialised.[18] The present concern is whether the general law can or should do anything about this. The Directive lacks even the vague screening standards provided by U.K. pollution control legislation.[19]

The second stage involves consultation focused upon the information supplied by the developer. Public authorities "likely to be concerned by the project by reason of their specific environmental responsibilities" must be consulted (Article 6). The public concerned must also be given an opportunity to express their opinion and the environmental statement must be made available to the public (Article 6.2). However it is for member states to determine who are 'the public concerned' and also the consultation arrangements (Article 6.3). There is evidence that notifications, consultations and information gathering are not carried out to a high standard.[20]

The final stage is where the Directive runs out of steam. The environmental statement, and the result of the consultation must be 'taken into account' in the development consent process. The "public concerned" must be informed of the decision and any conditions attached to it but need be told of the reasons and 'considerations' on which the decision is based only where local legislation so provides.

The Directive is silent as to the form the 'assessment' should take and as to the weight, if any, that should be given to the information collected. It makes no provision for monitoring the project should it be permitted to go ahead. Nor does it include any general enforceable right to disclosure of information. Indeed it preserves the freedom of national laws in relation to commercial and 'public interest' confidentiality (Article 10).

Under the Fourth Action Programme of the European Communities[21] public access to environmental information was made a priority and a directive has been issued .[22] This applies only to public authorities and to bodies under the control of public authorities. The government's consultation paper on the subject does not enthusiastically embrace openness. It emphasises the need for commercial and national security, confidentiality, the desirability of encouraging the 'voluntary' supply of information to public authorities, and the need to avoid bureaucratic burdens.[23] The government proposes no special machinery for enforcing disclosure of environmental information.

There is also the crucial issue of what projects must be subject to an environmental assessment. The Directive again gives considerable discretion to member states. The Directive distinguishes between two classes of project. In the case of those listed in Annex I there must always be an environmental assessment (Article 4.1). Annex 1 projects are large scale projects both public and private which necessarily have serious environmental effects. In the case of Annex 2 projects there must be an environmental assessment "where the member states consider that their characteristics so require" (Article 4.2).

It has been suggested that this entitles a member state to refuse to require any Annex II project to be subject to assessment but the U.K. government accepts that each Annex II project must be considered on its merits. The criterion is whether the particular project is likely to have a 'significant' effect on the environment by virtue of factors such as its nature size and location.[24] Member states can prescribe thresholds and standards (Article 4.2). They can also make rules subjecting general classes of development to assessment but can make exemptions only for specific cases. As we shall see the U.K. government takes a strict view of the meaning of 'significant'.

The alleged failing of the U.K.

In a Letter of Formal Notice received by the U.K. government in October 1991 the European Commission set out four general respects in which it alleged that the U.K. was in breach of the Directive. It also made seven specific allegations. Of the specific allegations, six were later withdrawn. In two cases this was due to further information received by the Commission, in one case to the project being withdrawn and in the other cases to a more satisfactory assessment subsequently being produced.[25] The issues of principle identified by the Commission therefore remain unresolved. They relate closely to English law's general attitude to environmental concepts and values.

These issues of principle are as follows:

1. The "pipeline" question. The Commission takes the view that projects for which application was made before the Directive came into effect but where consent was not received until afterwards should be subject to assessment.
2. The Commission is concerned about the exemptions from the need for planning permission, which are widespread in English law. These include much agricultural and forestry development, and all Crown development. The Commission takes the view that the Directive requires the U.K. to enact legislation imposing consent requirements in these cases. The U.K. takes the view that assessment applies only where a project requires consent already.[26]
3. The Commission takes the view that the U.K. is exercising too broad a discretion in deciding which projects should be subject to an assessment. According to the Commission the likelihood of

significant effects on the environment is an objective test whereas under some U.K. legislation the decision is left to the subjective judgement of designated authorities with a risk that the work will be carried out haphazardly and perfunctorily by relatively junior officials.

4. Most fundamentally, the Commission alleges that the U.K.'s method of implementing the Directive does not count as an environmental assessment at all but merely requires planning authorities to consider miscellaneous information in an unstructured way. The developer's environmental statement (ES) is itself a kind of assessment but under the Directive the assessment must be carried out by the "competent authority".

An 'assessment' should entail an independent weighing up of costs and benefits. This is certainly not the case in English law. In Spain for example a separate public body produces the environmental assessment which is then weighed against other factors by the body that consents to the project as a whole.

The Commission is proceeding with one specific allegation against the U.K. This is the Oxleas Wood case, where the adequacy of the information supplied to the public by the developer, the Department of Transport, is in issue. In other cases, notably the Sizewell inquiry, the most thorough environmental information has been provided by objectors.[27] This is therefore a major test of the efficacy of the Directive.

Enforceability of the directive

It is widely recognised that the European Community system is weak when it comes to applying Community Law at the national level.

Directives are addressed to member governments and require them to make the necessary implementing laws. They are not as such enforceable in the English Courts nor by individuals in the European Court. If a member state fails to comply with a directive the European Commission must institute proceedings in the European Court under Article 169 of the Treaty of Rome. This procedure has no enforceable sanction[28] and the Commission's ability to obtain information appears to be haphazard and unreliable. The obligation derived from Article 5 to provide effective remedies for breach of Community Law is binding upon English courts[29] but as we shall see the courts have concentrated on the concept of "rights" and "environmental rights" as such are difficult to fit into the categories of 'right' recognised by English law. English law has developed in terms of specific relationships and particular kinds of damage quantifiable in cash terms.

In *Frankovich v Italy*[30] the European Court of Justice held that failure to implement a directive could be enforceable against the government in a national court. It is doubtful whether this is itself of much help in relation to the Environmental Impact Assessment Directive. Damages under *Frankovich* are available only where the particular directive confers rights upon individuals the contents of which can be ascertained from the directive itself.

This is clearly not applicable to the Environmental Impact Assessment Directive which is concerned with public rights. A public law remedy would be more appropriate in the environmental sphere but in *Regina v Secretary of State for Employment ex parte Equal Opportunities Commission*[31] it was held that the courts could not issue mandamus or make a declaration to compel the Secretary of State to introduce amending legislation in order to implement an EC directive. This is a matter for Parliament alone. Furthermore, *Frankovich* could not be used against a local authority for failing to apply the Directive properly but only, probably, against the central government for failing to implement it.

Directives may sometimes be enforced directly in national courts against public authorities but not against private persons. A directive has "direct effect" in this sense i) if it is sufficiently clear, precise and unconditional not to require the exercise of a substantial amount of discretion by the member state in order to implement it; ii) if the date prescribed for implementing the directive has passed; and iii) if it confers rights upon the person seeking to enforce it.[32]

It is not clear to what extent the Environmental Impact Directive has direct effect. Firstly, much of the substance of the Directive concerns obligations imposed upon a private developer against whom the Directive cannot be enforced. Secondly, as regards i) specific requirements of the Directive leave considerable discretion to the local authority, notably the question whether a project falls within Annex II and also the content of an environmental statement. This kind of discretion does not make the Directive unclear, imprecise or unconditional as such, but concerns the application of a directive to an individual case as opposed to its implementation. There is no reason why matters of application should prevent a directive being of direct effect but the existence of discretion means that the issue would be justiciable only in public law judicial review proceedings. On the other hand the publicity arrangements contemplated by the Directive are left to member states so that the Directive is unlikely to be of direct effect in that respect. In the case of publicity the discretion is not merely one of application to individual cases but of deciding as a matter of legislative principle what kind of publicity arrangements should be made.[33]

Thirdly, as regards iii) it is not clear what is meant by a 'right' in this context. In *Twyford Parish Council v Secretary of State for the Environment*[34] it was held that the Directive was sufficiently clear, precise and unconditional to be of direct effect, but that there was also a requirement that the rights of the particular plaintiffs must have been prejudiced. In that case the applicants were a Parish council and representatives of pressure groups. They did not allege any specific harm other than a failure to carry out an adequate assessment and in particular a failure to supply an adequate non technical summary of the environmental information. Given that the purpose of environmental assessment is to provide information about the likelihood of future harm and that, environmental interests are essentially public interests, this further limitation upon direct effect is questionable. If *Twyford* is correct it means that the Directive is enforceable only in cases where private law rights have been infringed even though, as a matter of locus standi, pressure groups, local authorities and other persons without private rights are able to apply for judicial review. *Twyford* was a public law action to challenge a highways decision made by the Secretary of State. The absence of prejudice is a matter relevant to the court's discretion to quash and should not have been confused with the question of direct effect.[35]

There is no reason in principle why purely public law rights should not qualify for the purpose of direct effect. The essence of the direct effect doctrine is the principle of justiciability which in public law is subject to different considerations than in private law.[36] Indeed it is arguable that a failure to apply the directive lawfully is, in itself the violation of a public law right to an assessment and, subject to *locus standi*, fully justiciable in public law proceedings without further prejudice being required.[37]

Even where a directive is not of direct effect, domestic legislation implementing a Directive must be interpreted in the light of the Directive in accordance with the purposive methods of construction used by the European Court.[38] Moreover implementing legislation is *ultra vires* if it is inconsistent with the directive which it purports to implement. However in *Swale Borough Council ex parte R.S.P.B.*[39] Judge Simon Brown refused to do either and preferred a literal construction of the U.K. regulations. The case concerned the reclamation of mudflats in the Thames Estuary. The U.K. regulations (Schedule 2) appear to have misapplied the Directive since they limit the requirements for assessment only to agricultural development whereas Annex II of the Directive seems to include all land reclamation projects. In relation to the community obligations Judge Simon Brown remarked that the court is not entitled on judicial review "to reach its own decision so as to ensure that our EEC treaty obligations are properly discharged". This may be the case in relation to attempts directly to enforce the Treaty in the absence of direct effect[40] but in relation to the vires of implementing legislation the point seems irrelevant. In the particular case the limitation to agriculture could probably be struck out leaving the rest of the regulation intact.[41]

Suppose a non conforming regulation were to be made under the new power conferred by section.71A of the Town and Country Planning Act 1990.[42] While the lawfulness of the regulation could not be challenged if it went beyond European requirements it should arguably still be construed to conform to the Directive and would not prevail against a directly effective provision of the Directive.

English legal culture

The administration of the environmental assessment machinery depends upon discretionary decisions made by central and local government. The 'environmental information' comprising the developer's environmental statement coupled with the result of consultations and representations are simply taken into account as part of the ordinary planning process. Subject to directions by the Secretary of State, it is for the local authority to decide whether assessment is required at all and also the amount of information required. The same local authority and, on appeal, the Secretary of State decides whether or not to grant planning permission. In planning law as in other areas of governmental activity, wide administrative discretion is paramount. The authority can take into account any material consideration and these include social and economic considerations as well as environmental factors. It is also lawful to take into account political ideology, in favour for example of market forces.

The general law in the hands of the courts represents values, concepts and assumptions against which the operation of the assessment procedure can be evaluated. In a liberal democracy the courts should adopt a broadly neutral stance in relation to issues on which there is no consensus. Environmental impact assessment is not, as such, an environmental protection measure with positive goals. Environment impact assessment is intended to enable decision makers to make an informed choice between environmental and other objectives and for the public to be consulted. Nothing in the Directive requires any particular weight to be attributed to environmental factors or to public opinion.

On the face of it English law reflects this liberal ideology. The courts are deferential to executive discretion provided that it has some rational basis and does not violate what the court regards as widely accepted standards of morality.[43] On the other hand liberal principles do not require the courts to be deferential in respect of 'process' matters. The courts should be able to ensure that the information provided is sufficient, that the viewpoint of all interested parties is given equal opportunity and that an assessment is carried out in the manner contemplated by the Directive. In these respects English law does, arguably fall short. Firstly, the courts are deferential to discretions exercised by other bodies, particularly governmental bodies but also professionals including possibly environmental consultants. Secondly, the courts do not recognise environmental interests as such as "rights". The neutrality principle breaks down at this point because property and financial interests weigh more heavily in the scales than the concerns of the environmental lobby. Thirdly, the literalistic way in which the English courts interpret the Directive may run counter to its purposes.

Deference

The court's deferential philosophy is manifested in ways which are not conducive to the aims of the Directive.

i) Consistently with the neutrality principle, the court will not attempt to weigh the competing considerations or evidence that is taken into account. Provided that the authority considers a relevant factor it can attach little or even no weight to it. For example in the case of *Regina v Poole Borough Council ex parte Beebee*[44] even though the land concerned was in an especially important Site of Special Scientific Interest the weight to be attached to this was solely for the authority to decide. Where special weighting is required it is necessary for the government either to legislate or to publish policy guidance which can create presumptions in favour of a particular factor. The courts then attempt to follow the priorities so expressed. There is now a general statutory presumption in favour of the development plan so that environmental impact considerations can be given special weight by incorporation into the plan.[45] The courts alone have no power to privilege environmental concerns so that criticism of the Poole case on this point is misconceived.

ii) The position is different in respect of the adequacy of the environmental impact assessment as a fact finding process. Here the courts do have the power to ensure that the information provided is adequate. In other contexts the courts have held that consultation is a two way process and that sufficient time and information must be provided for a genuine interchange.[46] The rules of natural justice are based upon the proposition that both sides must be given an adequate hearing. In the USA the courts have developed a "hard look" doctrine in the context of environmental issues. This requires that an environmental impact assessment should adequately consider relevant alternatives, that all interested parties should have access to the process and that a proper cost/benefit analysis should be carried out.[47]

The English courts have rejected any cost/benefit requirement and have usually refused to take a hard look.[48] Firstly, the rules of natural justice have concentrated upon the protection of individual rights and not upon participation in the decision making process as such.[49] Secondly, the courts will interfere with findings of fact only where there is no supporting evidence at all. They will not attempt to quantify, evaluate or structure evidence. The cases concerning the Directive have consistently upheld this approach. In the *Poole*[50] case which concerned a proposal to develop housing in a Site of Special Scientific Interest, no formal environmental impact assessment had been carried out at all. The council, who were themselves the developers had not apparently considered whether or not to do so. Nevertheless they had possibly considered some relevant environmental data and Judge Schiemann refused to quash the planning permission. (The Secretary of State subsequently stepped in and did so). His Lordship said: "The authority had in their possession the substance of what they would have had if they had applied their minds to the 1988 regulations. The substance of all the environmental information which was likely to emerge by going through the formal process had already emerged and was apparently present in the council's mind". This ignores the crucial participatory aspects of environmental assessment and arguably violates the principle of European law that domestic law cannot be relied upon contrary to a directive.[51]

Similarly in Twyford Down,[52] Judge McCulloch held that the court would interfere only if "what was provided could not be regarded by anyone in the position of the Secretary of State as sufficient for this purpose".

Broader questions of fairness relating to the disclosure of traffic forecasts also arose in that case. Again the court refused to intervene because of the lack of harm[53] to the particular applicants. The question of the "unrepresented third party" i.e. the public appears to have been treated as a matter in the custody of the Secretary of State rather than as one for direct procedural recognition.

These cases concerned the adequacy of the assessment carried out by the public authority and were pursued by objectors to the scheme. There is no reason to suppose that a more stringent standard would be applied to the environmental statement prepared by a private developer.[54] In other contexts it has been held that the methodology used by the Department of Transport to prepare traffic forecasts is not open to question in the courts.[55] Local authorities are not generally obliged to seek out alternative sites[56] and matters which are regarded as matters of planning judgement, for example compatibility with the locality are not challengeable.[57]

However where the prediction of adverse consequences was a factor against a proposed development a stricter approach seems to be taken. The courts have required a prediction against the development to be justified by a balance of probabilities test. If this applied in the environmental assessment context it would tip the balance in favour of the developer's environmental statement.[58]

These restrictions upon judicial control go to the heart of the environmental impact assessment process. Judge Schiemann's argument in Poole was essentially the same as that urged by the U.K. government prior to the introduction of the Directive, namely that existing discretionary mechanisms were adequate. This ignores the scoping and participatory elements in the assessment process as well as begging the question of what an 'assessment' is meant to be. The loose legal regime is conducive to haphazard practise and low priority.

The Directive distinguishes between two levels of environmental information. Article 5.2 prescribes basic minimum information about the project itself, the measures envisaged to combat adverse environmental affects and the data required to identify and assess environmental effects. This must always be included in the Environmental Statement. A non-technical summary is also required. Annex III requires more sophisticated information about alternatives, supporting reasons, the production process, forecasting and technical problems in compiling information. According to Article 5(1) of the Directive the second category must be included "in as much as (a) the member states consider it relevant to a given stage of the consent procedure and to the specific characteristics of a particular project or type of project and of the environmental features likely to be affected; (b) the member states consider that a developer may reasonably be required to compile this information having regard *inter alia* to current knowledge and methods of assessment".

This could be interpreted broadly as imposing a general requirement to include Annex III data subject to the right of the member state to relax that requirement in particular circumstances. However the U.K. regulations adopt the opposite approach. By virtue of Schedule 3, the minimum information required under Article. 5(2) must be provided but the more sophisticated Annex III information is made voluntary, subject to a power under regulation 21 to require further information.

iii) The same deference applies to the decision whether to require assessment at all. It could be argued that the definition of the various categories of project specified in the Directive is a matter of law for the courts. The courts have not taken this approach but have treated the matter as one of fact and judgement, thus attracting only a minimal level of review. In *Swale*,[59] Judge Simon Brown said that the question of whether an assessment is required is "exclusively for the planning authority subject only to Wednesbury challenge" (i.e. challenge for complete irrationality).

iv)The onus is on the person who seeks to challenge an environmental assessment decision to show that the decision is flawed, for example by virtue of failing to take environmental information into account. The court will not probe into the facts and will usually look only at the formal documents provided by the decision maker. For example in the *Poole* case (above) the issue was whether there had been a proper consideration of the impact of a proposed housing development upon an important Site of Special Scientific Interest which was a haven for several rare species. In the absence of contrary evidence, and even though it acknowledged the possible unfairness involved, the court felt bound to assume that the competing factors had indeed been considered.

v) Even where it has been established that a public authority has acted unlawfully, the court has a discretion not to set aside the offending action or to give the citizen a remedy. It will exercise this discretion if it considers that the citizen's interests have not been sufficiently injured or on the grounds of administrative convenience. In the Poole case for example Judge Schiemann thought that carrying out a proper assessment would be pointless in terms of what it would uncover and in *Swale*[60] Judge Simon Brown thought that the applicants had been guilty of undue delay in not bringing their action until very near the expiry of the three months time limit. He also thought that the procedural right to be consulted was outweighed by financial loss that would be suffered by contractors if the planning permission was quashed.

Private rights

There is an underlying bias in favour of financial interests. This was formed in the laissez faire era of the nineteenth century upon the foundations of property rights and freedom of contract and was incorporated by judicial interpretation into the post World War II planning system.[61] Within this framework, environmental interests are not recognised as 'rights' and must therefore be protected, if at all, by specific legislation. This is fundamental to the Directive which is drafted upon the basis of abstract environmental concerns.

The absence of environmental rights reveals itself in several ways. I have already mentioned the emphasis on private rights in relation to the direct effect doctrine and the evidential burden in favour of the developer when it comes to predictions of adverse consequences. Other examples are as follows:

i) Subject to the development plan there is a general presumption in favour of development.[62] Therefore the burden of proof (albeit not a strict legal burden of proof) lies upon those who assert environmental interests. Local authorities are free to adopt other presumptions which might outweigh the presumption in favour of the developer.

ii) Authorities are required to give reasons for anti development decisions but not for pro development decisions. Thus reasons must be given for a decision to require an environmental impact assessment but not for a decision not to require one.[63] Similarly reasons must be given for refusing planning permission or for imposing conditions but not for granting planning permission. Pro-development decisions are therefore relatively difficult to challenge. This has coloured the courts' approach to environmental challenges. For example in the *Poole* case Judge Schiemann said that "it seems to be the consensus view that society would rather live with the possible injustices of not knowing the reasoning of the decision making body than with the inconvenience of forcing it to articulate reasons."[64]

iii) There is a right of appeal to the Secretary of State with arrangements for a public inquiry where planning permission is refused but not where permission is granted. Indeed only the applicant for permission can appeal. Third parties who wish to challenge a planning permission on environmental grounds must therefore resort to the ordinary courts thus facing procedural and financial barriers.

iv) There is no public right to challenge a decision in the ordinary courts. Environmental interests in themselves are not enough. For example in the *Rose Theatre* case a pressure group formed to attack an office development proposal that affected a unique fragment of Shakespearean theatre was denied standing.[65] The courts do however appear to be sympathetic towards well established pressure groups, particulary those with government backing and especially where they have invested substantially in the issues in financial terms. For example in the *Poole* case, the British Herpetological Society was given standing but not the World Wildlife Fund in its own right.[66] In *Lewin*[67] the Society for the Prevention of the Field of the Battle of Naseby was given standing. There was no discussion of the issue. However the Society had been involved with the issues for a considerable time and had participated in lengthy public inquiries.

v) The purpose of citizen participation at a public inquiry is regarded by the courts as essentially to ventilate individual grievances and not to represent general policy concerns. For example in *Bushell v Secretary of State* [68] citizen participators representing environmental concerns in a motorway inquiry wished to cross examine civil servants about the government's methods of forecasting traffic demands. The House of Lords upheld the government's refusal to expose this material on the ground that these were matters of policy and confidential within the department.

It will be recalled in this connection that the Directive requires public consultation. Indeed the relevant Commissioner is on record as saying "I am inflexible on this point. I will not accept any limitation on the Commission's rules of public consultation. Every citizen has a right to be heard on the decisions that will affect their environment and thus their life".[69] Nevertheless the Directive permits member states to decide what is meant by the 'public' and the manner in which the public is to be consulted (Article 3). The English regulations require that a copy of the environmental statement be available for public inspection, that individuals may buy a copy "so long as stocks last", and that where a planning application is accompanied by an environmental statement the application be advertised for twenty eight days in a local newspaper and in a site notice.[70] However failure to advertise does not necessarily make the planning application invalid. The Department of the Environment circular addressed to local authorities which purports to explain and advise upon the regulations is content to describe the legal arrangements without stressing the importance or otherwise of public consultation. The circular makes it clear that any additional publicity e.g. exhibitions, or separately publishing the non-technical summary required by the Directive is optional.[71]

vi) The court's preference for financially quantifiable interests means that when the court is balancing competing factors, environmental interests carry little weight. This is important because judicial review remedies are discretionary so that victory on the merits does not guarantee a favourable decision. In *Swale*[72] Judge Simon Browne refused to quash the planning permission on the ground (among others) that the financial interests of the developer and its contractors outweighed the environmental concerns of the applicants which were at most a right to be consulted. It must be admitted however that the decision in *Swale* was coloured by the fact that the applicants had been guilty of 'undue delay' within section .31(6)b of the Supreme Court 1981. This entitled the court to consider whether granting relief would "substantially prejudice the rights of any person". On the other hand his Lordship did not appear to regard the environmental interests of the public as 'rights' within the equation. Neither in *Swale* nor in *Twyford*[73] were the applicants regarded as representing the public interest.

Semantic textual analysis

A third characteristic of English legal culture is a semantic and literalistic approach to drafting and interpreting legal rules. It would be an oversimplification to contrast this automatically with a purposive approach because some purposes may well be advanced by semantic techniques. However in the context of European law the semantic approach is especially likely to clash with the policies of the European Community because their policies are implemented by draftsmen schooled in the more open textured and purposive methods used by continental lawyers. Many of the allegations made by the Commission against the U.K. turn on this kind of conflict.

i) According to the Commission all projects falling within the designated classes should be subject to assessment. The English regulations by contrast are drafted on the basis that only projects which under the existing law require specific planning permission are subject to the Directive. There are several important exemptions from the need for planning permission. These include development by the Crown the specific subject of a Commission allegation now withdrawn; agriculture and forestry; development within existing special planning zones and enterprise zones, and certain classes of development specified by ministerial order as 'permitted development'. The Commission has alleged that several kinds of project falling within these exemptions require assessment.

Article 2 of the Directive states that the assessment procedure may be integrated into the existing consent procedures and states that "before consent is given projects likely to have significant environmental effects should be subject to an assessment". A literal construction therefore probably supports the U.K. government whereas a purposive construction would not do so.

ii) A related issue is the meaning of "project". The Directive defines project broadly to include the whole enterprise that is proposed (Article 1). The U.K. government appears to equate "project" with the particular application for planning permission. The scope of this can be determined by the developer himself.

Planning permission in English law does not relate to projects as such but to the artificial concept of development. Conceptually development centres upon processes rather than end products. Any 'project' can be broken up into separate components each consisting of one or more separate acts of development i.e. building, engineering, mining or other operations or material changes of use.[74] The High Court upheld the government's approach in the *Swale* case[75] by stating that "the question must be answered strictly in relation to the development applied for, not any development contemplated beyond that. A large project that as a whole might require assessment could therefore be broken up into separate smaller applications thus escaping the regulations. *Swale* involved a land reclamation application on mudflats which were an important area for water birds. This was part of a larger scheme involving a marina and other forms of commercial development. Taken as a whole the project would arguably require assessment. Taken in isolation the land reclamation scheme would not.

The same approach was taken in *Lewin v Secretary of State*[76] which concerned the M1-A1 link road. It was held that a side road order could be treated as a separate project and therefore fell only within Annex II of the Directive. Treating the side road as part of the main road project would have made environmental assessment compulsory under Annex I.

The Commission has dropped allegations in relation to the M11 link road, the Channel Tunnel Kings Cross Terminal and a factory in Northamptonshire. All of these were alleged to involve the artificial splitting of a complex project into smaller components.

iii) The Commission takes the view that any project which falls within the Directive must be subject to assessment where the final consent was not given by the date on which the Directive took effect (3 July 1988). On this view, projects which were in the pipeline on that date must be held up for an assessment. However, the English courts have held that the Directive does not apply to pipeline projects.[77] Unusually the court adopted a broad interpretation of Article 2. However this owes less to the purpose of the Directive and more to traditional values of deference to government and to commercial interests in this case reinforcing each other. The administrative complications that would result if the Directive were to apply to pipeline projects were used as a way of distinguishing earlier European cases[78] because the environmental assessment procedure is applicable at every stage of the planning process.

iv) Finally there is the crucial term "significant". An Annex II project is not subject to assessment unless the member state considers that it is likely to have a 'significant' effect on the environment (Article 2.1, Article 4.2). A broad meaning of 'significant' would be 'of significance' - something not merely trivial or insubstantial. The U.K. Government has taken a narrow approach and appears to regard 'significant' as meaning "unusual" or "exceptional". In Circular 15/88 paragraph 20 for example the Department of the Environment suggested that environmental impact assessment would, in the case of Annex II projects, be required only in three main types of case: (i) for major projects of more than local importance; (ii) 'occasionally' for projects on a smaller scale which are proposed for particularly sensitive or vulnerable locations; and (iii) 'in a small number of cases' for projects with unusually complex and potentially adverse environmental projects. The circular also opines that the number of projects falling within these categories will be a small proportion of all Schedule II projects and states that public opposition or controversy is not in itself relevant.

This specific point has not reached the courts. However the *Swale* case indicates that the courts would not intervene. Judge Simon Brown said that the decision whether or not any particular development is or is not within the scheduled descriptions is exclusively for the planning authority subject only to *Wednesbury* challenge. In the context of a measure implementing a European Community. Directive this is probably an inadeqüate threshold of review.

Conclusion

The framework of English law gives local authorities and central government considerable discretion as to whether to implement the Environmental Impact Assessment Directive broadly or narrowly. The courts have remained faithful to traditional concepts of deference to executive discretion, to preference for financial interests and to a literal reading of domestic rules without reference to the objectives of the Directive. They seem to have ignored the obligation to construe implementing measures purposively or at least to have assumed that this obligation does not apply to judicial review. This approach is facilitated by the absence in English law of a notion of environmental or public rights. The Directive has been absorbed into the traditional framework without apparent recognition of the distinctive nature of environmental concerns or of the participatory character of environmental impact assessment.

The Directive itself is often vague and evasive, notably in relation to what an environmental assessment comprises and to issues of public participation. One of the functions of the European Commission is to act as advocate for European interests and it is fulfilling this function by taking proceedings against the U.K. for allegedly failing to implement to Directive correctly. What is missing is firstly, independent machinery for monitoring the environmental assessment process and secondly, adequate support for environmental interests.

Uniformly unsuccessful legal challenges have concerned governmental decisions either not to require assessment at all, or to be content with minimal levels of assessment. There is no evidence of dissatisfaction on the part of developers with overzealous, environmentally concerned local authorities. One of the main advantages proposed for the Directive has been that it would focus attention upon environmental issues and place them in the centre of the planning stage.[79] English law has made little contribution to this objective. Wood and Jones[80] recommend imposing consultation duties prior to the preparation of an Environmental Statement, requiring the authority rather than the developer to prepare the Statement and setting up an independent review body.

This could be taken further. It would be desirable to create an independent monitoring body with wide powers along the lines pioneered by the Council on Tribunals. This body could monitor practise and advise upon procedural standards. It could also commission research and possibly investigate complaints from the public. It could liaise between the European Commission and the U.K. environmental assessment community.

Such an independent body might also exercise the quasi judicial function, which at present belongs to the Secretary of State, of giving directions under regulation 6 as to whether a proposed development in Schedule 2 is likely to have 'significant effects on the environment'. Where the local planning authority has given an opinion in writing to this effect under regulation 5, the applicant may apply to the Secretary of State for a direction - essentially a right of appeal. This power to decide whether a project should be subject to assessment is not appropriate for political resolution but nor is it sufficiently objective to be treated as a question of law. An independent panel of the kind suggested here may be a solution.

The watchwords of the Franks Committee, whose report in 1958[81] led to the creation of the Council on Tribunals, were "openness, fairness and impartiality". The watchwords of the environmental impact assessment panel might be "scoping, screening and participation". As the law stands, the U.K.'s environmental impact procedures fail to ensure compliance with these basic techniques.

As regards support for environmental interests, the main cases which I have discussed in this chapter were not appealed. This may be partly due to the fact that environmental groups have limited money and little access to legal aid whereas developers can often absorb legal costs easily. In a regime where "costs follow the even", this places environmental interests at a considerable disadvantage. There are various ways in which the costs rules could be changed so that the expense of public interest litigation is more fairly shared.[82] This applies to public inquiries as well as to litigation.

Most fundamentally, the deferential character of English legal culture is now being challenged by European values which give greater emphasis to environmental rights than can apparently be accommodated within English law.

Footnotes

1. European Community Directive no 85/337 (OJNOL 175/40 1985 p5)
2. See 11th Report from the House of Lords Select Committee on the European Committess 1980-81: *Environmental Assessment of Projects,* House of Lords Paper 69. Delay appears not to be a problem in practice. See Kenyon (1990) *JPL* 419.
3. EC Directive 85/337 preamble. See also Williams 20 *Planning Outlook* p54 (1983); Ball and Bell, *Environmental Law* (1991) Ch 4; OECD, *The Polluter Pays Principle* (1975).
4. For example, *Hall v Shoreham UDC* (1964) 1 All ER 1; *R v Hillingdon BC ex parte Royco Homes Ltd* (1974) QB 720. This is also reflected by the notion that planning gain cannot be required except to compensate particular planning disadvantages of the proposed application. See Purdue Healey and Emmis Healey (1992) *JPL* 1012. This view seems to be based upon the belief that planning is not supposed to improve the environment but merely to maintain the *status quo.*
5. See *St Helens Smelting Co Ltd v Tipping* (1865) 11 HLC 642; *Halsey v Esso Petroleum* (1961) 1 WLR 683. Government policies are however relevant in nuisance law in relation to identifying the environmental standards of the area. *Gillingham BC v Medway Docks Ltd* (1992) 3 All ER 923.
6. See *Haywood v Brunswick Permanent Building Society* (1881) 8 QBD 403.
7. cf *Wrotham Park Estates v Parkside Homes* (1974) 1 WLR 798, amenity without economic value recognised although within a proprietary framework.
8. Articles 130R, 130S, 130T.
9. Article 100a now requires a 'high level of protection' for environmental standards. These provisions require uniformity of standards within member states. Environmental considerations can justify an interference with the operation of the common market: See *Commission v Denmark* (1989) 50 *CMLR* 619.

10. See also Town and Country Planning Act 1971 section.71A which allows environmental assessment to be required beyond the scope of European requirements. See *ENDS* Report 216, (1993) p.33.
11. See Salter (1992) *JPL* 313.
12. *Marleasing SA v La Commercial Internacional de Alimentacion SA* (1990) *ECR* 1-4135. In *Duke v GEC Reliance Ltd* (1988) AC 618 the House of Lords had limited this to cases of (i) directives with direct effect (below); (ii) legislation implementing specific directives. See Docksey and Fitzpatrick (1991) 1 *LJ* 113, de Burca (1992) 55 *MLR* 215. See also *Von Colson v Land Nordrhein-Westfalen* (1986) 2 *CMLR* 430; *Lister v Forth Dry Dock and Engineering Co Ltd* (1990) 1 AC 546; *Finnegan v Clowney Youth Training Programme Ltd* (1990) 2 AC 407. The often expressed view that *Lister* and *Finnegan* concern Parliamentary Supremacy seems misconceived since both concerned subordinate instruments. Whether or not all legislation which implements a directive must be interpreted to conform to the directive even where it is not ambiguous, (cf *Marshall v Southampton and West Hampshire Area Health Authority (Teaching)* 1991 1 CR 136), implementing subordinate legislation must be so interpreted as in the case of the Environmental Impact Directive (Lister above).
13. See Mertz (1989) *JPL* 483; National Environmental Policy Act 1969; Callies and Freilich, *Cases and Materials on Land Use* (1987); the main public authority provisions are:
The Environmental Assessment (afforestation) Regulations 1988 (SI No 1207); The Highways (Assessment of Environmental Effects) Regulations 1988 (SI No 1241); The Electricity and Pipe-Line Works (Assessment of Environmental Effects Regulations 1990 (SI No 442); The Harbour Works Assessment of Environmental Effects (Regulations) 1988 (SI No 1336); The Harbour Works (Assessment of Environmental Effects) (No 2) Regulations 1989 (SI No 424); The Land Drainage Improvement Works (Assessment of Environmental Effects Regulations 1988) (SI No. 1217);
The Directive does not apply to projects which require approval by Act of Parliament (Article 1 para 5) although the phrase "since the objectives of this Directive are achieved through the legislative process" may be a mistranslation of 'provided that'.
In practice environmental assessments are made for many such projects and Parliamentary procedures have been revised.
14. Art 13.
15. Town and Country Planning (Assessment of Environmental Effects) Regulations 1988 (SI 1988 No 1199), hereinafter 'regs'.
16. Direct and indirect effects on human beings flora and fauna, soil, water, air, climate, landscape, material assets and cultural heritage.
Curiously, while the interrelation between natural phenomena must be considered the interrelation between these and 'material assets and the cultural heritage' need not.

17. Directive Article 5(3); Regs 8(4), 22.
18. See Wood and Jones, *Monitoring Environmental Assessment and Planning.* (Department of Environment 1991), paras 5.12 - 5.22.
19. Environmental Protection Act 1990, s.7, s.8(7) (BATNEEC BPEO)
20. *Op cit.* note 18, para 5.6. The Government has undertaken to extend consultation rights to supplementary information. See ENDS Report, 216 p33.
21. (1987) OJ C70 March 18 1987 p3.
22. Directive 90/313/EEC. Deadline for compliance 31.12.92
23. HMSO January 1992. See *Water Law*, May 1992 p63.
24. Reg 2. Directive, Preamble, Art 1.1, Art 2.1
25. See (1992) *JPEL* 914.
26. However the UK is considering the extension of assessment to certain large scale agricultural and forestry projects. Forestry projects supported by government grant already require assessment. See S.I 1988 No 1207 (above note 13). See 1993 216 *ENDS* Report 33.
27. See Purdue (1989) 4 *Connecticut Journal of International Law*, 387.
28. Art. G (51) of the Treaty on European Union (Maastricht) gives the European Court of Justice power to award a lump sum or penalty payment against a state that fails to fulfil a Treaty obligation within a time limit laid down by the Commission.
29. See *Commission v Greece* (1989) ECR 296; *Simmenthal* (1978) ECR 629; *Factortame Ltd v Secretary of State for Transport (No 2)* (1991) 1 All ER 70. Article 5 requires member states "to take all appropriate measures ... to ensure fulfilment of the obligations ... resulting from action taken by the institutions of the community." 'Rights' are not necessarily required.
30. Cases 6/90, 9/90 *The Times,* Nov 20 1991. See (1992) 55 MLR 690.
31. (1993) 1 All ER 1022.
32. See Kramer (1991) *JEL* 914.
33. See Art. 3.
34. (1992) JEL 274 cf *Becker v Finzanzamt Munster-Innenstadt* (1982) ECR 53; *Marshall v Southampton Area Health Authority* (1986) ECR 723. See Macrory (1992) *JEL* 273.
35. *Miller v Wymouth and Melcombe Regis DC*, 27 P & CR and CR 468. See also *Kincardine and Deeside DC v Forestry Commissioners* (1991) cited Salter (1992) JPL 14 at 15.
36. See *Gouriet v Union of Post Office Workers* (1978) AC 435.
37. cf *R v Secretary of State for Employment ex parte EOL* (1993) 1 All ER 1022 – this turns on the special statutory position of the EOC.
38. *Op cit.* note 12 and associated text.
39. (1991) *JEL* 135.
40. See *R v Secretary of State for Employment* (above note 37); *Brind v Secretary of State for Home Affairs* (1991) 1 All ER 720.
41. See *DPP v Hutchison* (1989) 1 All ER 1060.
42. *Op cit.* note 10.

43. *Council of Civil Service Unions v Minister for the Civil Service* (1984) 3 All ER 935 at 951.
44. (1991) *JPEL* 643.
45. See Town and Country Planning Act 1990 section 54A. However this probably affects only the onus of persuasion and the courts will still not interfere with matters of weight except in extreme cases, eg the authority could lawfully decide that the presumption in favour of granting permission outweighs a presumption in favour of giving 1effect to an adverse assessment built into the plan unless the evidence was overwhelming. See eg *Devon CC v Secretary of State* (1993) JPL 40; *St Albans District Council v Secretary of State for the Environment* (unreported 23 November 1992).
46. See eg *R v Secretary of State for Social Services ex parte AMA* (1986) 1 WLR 1.
47. See Craig, *Administrative Law* (2nd ed) pp300-303. *County of Iryo v City of Los Angeles* 71 Cal App 3d 185, (1977) – lack of structure – alternatives not canvassed. *Whitman v Board of Supervisors* 88 Cal App 3d 397 (1979) – cumulative impact not examined. cf Sax The (Unhappy) Truth about NEPA, 26 Okla L Rev 239 (1973). Bordeck and Pugliamesi, The Environmental Impact Statement v The Real World, Public Interest, No 49, at 22, 24, 36-37 (1977).
48. See *London Residuary Body v Secretary of State* (1990) 2 All ER 309
49. See *Bushell v Secretary of State* (1981) AC 75; *George v Secretary of State* 38 P & CR 609.
50. *Op cit* note 44. See also *Lewin v Secretary of State* (1992) JPL 342.
51. See Pescatore (1983) 8 ELR 155, 189; Ratti (1979) ELR 1629 at 1642.
52. Above note 34.
53. See *Bushell v Secretary of State* (1981) AC 75 at 102.
54. The quality of these has, predictably, turned out to be uneven. Wood and Jones *op cit* note 18, pp 432-453, 5.14 - 5.19.
55. *Bushell v Secretary of State* (above note 53).
56. *GLC v Secretary of State* (1986) *JPL* 193; *Vale of Glamorgan BC v Secretary of State* (1986), 52P & CR 418.
57. *Luke v Minister of Housing and Local Government* (1968) 1 QB 172; *Stewart v Secretary of State* (1991) *JPEL* 103 – although reasons have to be given in general terms.
58. See *Westminster CC v Secretary of State* (1984) 3 All ER 737 at 742; *Vikoma International v Secretary of State* (1987) *JPL* 38; *Safeway v Secretary of State* (1992) *JPL* 565. However the test must be applied by the authority and is not an evidential burden in the reviewing court which will concern itself only with the question whether the proper approach was taken.
59. (1991) *JEL* 135.
60. *Ibid.* See also *R v Monopolies and Merger Commission ex parte Argyle Group plc* (1986) 1 WLR 763; *R v Secretary of State for Social Services ex parte AMA* (1986) 1 WLR 1.

61. See *Mixnams Properties v Chertsey UDC* (1965) AC 735; *Hall v Shoreham-by-Sea UDC* (1964) 1 WLR 240; *Buxton v Minister of Housing* (1961) 1 QB 278.
62. See *London Residuary Body v Secretary of State* (above note 48).
63. Reg 5(4); Reg 6(5)
64. *Op cit.*,, note 44.
65. (1990) 1 QB 504.
66. *Op cit.* note 44. See Hough (1992) JPL 342.
67. *Op cit* ,,note 44. See Geddes (1992) *JEL* 29.
68. *Op cit.* note 49.
69. See Salter (1992) *JPL* 214 at 218.
70. Regs 12, 13, 14 and 15.
    Any opinions given by the authority as to the need for assessment and any direction by the Secretary of State must also be available for public inspection (Reg 5(5) Reg 7) and, under the ordinary law copies of the application for planning permission and any grant of permission must be registered in a public register.
71. Circular 15/88 para 38; Appendix B para 42, 43 and 44.
72. *Op cit* note 38.
73. *Op cit* note 34.
74. See eg *Thomas (David) v Penybont DC* (1972) 1 WLR 1526; *West Bowers Farm Products v Secretary of State* (1985) JPL 857. However a dominant use test may be applied. See *Northavon DC v Secretary of State* (1980) 40 P & CR 32.
75. *Op cit.* note 38. Where the application does count as a project requiring assessment then of course the whole scheme must be considered in order to determine the environmental effects. *(Ibid)*
76. (1992) *JPL* 342.
77. Twyford Parish Council (above note 34). In *Bund Naturschutz and Others v Freistaat Baye,* ECJ Note of Proceedings Summary 32-92, 13 November - 4 December 1992, litigation has been commenced in the ECJ on this point.
78. *Becker v Finanzgamt Munster-Innestadt* (1982) ECR 53 at 76; *Marshall v Southampton Area Health Authority* (1986) ECR 1601; *Dik and Amor v College non Burgemeester en Wethouders Arnhem* (1988) ECR 1601.
79. See Carnwath (1991) *JEL* 57; Mertz (1989) *JPEL* 483; Grant (1989) 4 *Connecticut Journal of International Law* 463-477; Grant (1990) *Foro Administrative* 1104.
80. *Op cit.* note 17, para 5.22.
81. Cmnd. 218.
82. See Grosz, *TheGuardian*, 2 March 1993.

5 The Danish experience of clean technology: Co-optimising environmental protection and working conditions

Sussi Handberg, Per Christensen and Susse Georg

Introduction

The two most noted forms of social regulation — occupational health and safety and environmental regulation — are in Denmark, as in most other countries, distinct spheres of government policy. Legislation, administrative structures and enforcement procedures differ. Even the actors involved in the regulatory processes are different. This "division of labour" has its drawbacks, and there is a growing interest in integrating these two policy areas (Arbejdsministeriet & Miljøministeriet, 1990). There are two reasons for this. Firstly, many of the substances which create occupational hazards are also emitted to the environment as pollutants, e.g. improving ventilation to reduce hazardous fumes at the workplace can increase air pollution locally. Secondly, streamlining policies to improve the general environment and integrating them with policies to improve the working environment may help curb the public sector's growth, and perhaps improving government performance.

With the publication of the Brundtland Report sustainability has become a central issue in environmental policy — in Denmark as in many other European countries (World Commission on Environment and Development, 1987) — though not much emphasis in the Report is given to improving the working environment. The development of clean technology has an important role to play in achieving both goals: clean jobs and sustainability. Clean technology can avoid the "problem displacement," mentioned above, and/or the transferrence of pollution problems from one medium to another as is the case with "end-of-pipe" technologies. Seeking to reduce pollution at its source, clean technology goes to the core of how firms "do business". Managed correctly, implementing clean technology should — at least ideally — allow firms to simultaneously improve working conditions as well as environmental quality.

This chapter examines whether this indeed has been the case, by analyzing the regulatory approach of the Danish government and how industry has responded to these regulatory demands. The analysis is based on a case study of the regulatory histories of 16 firms in the fish processing industry in one specific region of Denmark (North Jutland.) Our case study indicates that these two policies have not been coordinated, let alone integrated, and that the development and diffusion of clean technology is largely the domain of the environmental authorities. Furthermore, our studies indicate that representatives from the firms' health and safety committees have been included late, if at all in the development processes associated with the implementation of clean technology. Labour has had very little influence on the development of clean technology in the fish processing industry.

Differences in regulatory approach

Environmental regulation is to a large extent based on "command and control" procedures, notably those associated with the pollution permits issued by the regional and municipal authorities. Usually the permits specify the emission standards with which the firms must comply, giving due consideration to the firms' economic and technological capabilities as well as the prevailing ecological conditions. The regional and municipal authorities have considerable discretionary power with respect to how these regulatory demands are formulated and enforced. Firms are for instance rarely prosecuted for non-compliance. Instead, the authorities rely on extra-legal sanctions (bargaining and/or advice) to achieve compliance with the regulatory demands.

Health and safety regulation is also based on a "command and control" approach, i.e. where government authorities — inspectorates in the various counties — monitor firm compliance with the national standards set by the Labour Inspectorate. The regulatory approach is, however, also based on a participatory model. The firms, i.e. their joint management-labour health and safety committees, are responsible for administering the health and safety law, while the inspectorate's role is to control the procedures and advise the firms on how to best comply with the law. The inspectorates can order firms to make specific improvements in the physical and psychological working conditions, but unlike the environmental authorities they do not approve of or issue permits for specific production processes.

While environmental regulation has mainly focused on reducing emissions, health and safety regulation seeks to eliminate the harmful consequences of work processes on humans. The health and safety law just like the environmental protection act, is a framework law. Only in this case, due to the way industrial relations are organized in Denmark, the parties on the labour market have quite a bit of direct influence on decisions about which occupational health and safety problems are to be regulated. One implication of these negotiations between the (central) government and the parties on the labor market, is that if the working conditions are to be improved, then this will have to be within the confines of the Collective Agreement and the

Agreement of Collaboration.

The development and diffusion of clean technology represents a challenge to the way in which environmental and health and safety regulation is to be undertaken in the future and to the way in which firms respond to these regulatory demands. First, one can expect a greater degree of overlap between these two realms of regulation, as environmental regulation will address changes in the production processes and not just emission reductions. Secondly, the firms' joint management-labour health and safety committees should — at least in principle — discuss these changes. Thirdly, the development of clean technology will presumably require an "up-grading" of worker qualifications through courses etc. The cooperative committees (which can also include specific technology committees) should be involved in deciding what educational measures should be taken in the light of the proposed changes in technology. The cooperative committees provide a structure within which employees can examine and discuss the technological changes which employers intend to introduce. This allows worker representatives and management to come together regularly, making it easier to anticipate and resolve conflicts.

The fish processing industry is dirty business

The most common *environmental problems* associated with the fish processing industry are the large amounts of process water used and the discharge of organic materials, nitrogen and phosphorous. The fish processing industries have normally discharged their untreated waste water to either a municipal waste treatment plant or to the local recipient. Difficulties in achieving ambient water quality prompted the Danish parliament to adopt an "Action Plan for the Aquatic Environment" in 1987. According to this plan, industry is required to use "best available technology", with due consideration being given to the economic capability of the firms.

The mobilization of public opinion, steadily rising water prices and user fees for waste water discharges have, along with the increasingly stringent emission standards, "forced" firms to use a wide array of cleaning and cleaner technologies. The former involves waste treatment measures such as flotation, separators and rotosieves. The latter has entailed optimizing the existing production processes through a variety of better house-keeping techniques (e.g. adjusting machines, improving maintenance etc.), increasing the internal re-use of water and making process improvements such as installing valve closures, re-fitting the nozzles of the filet-machines and developing a system to transport wastes without the use of water. These process improvements have enabled firms to reduce the discharge of organic material, nitrogen and phosphorous by 25-40%.

The most pervasive occupational hazards in these industries are the problems associated with the:

- humidity, temperature, draught and noise
- repetitive, monotonous work

Several studies from the early eighties have shown that work in the fish processing industry is associated with workers fatigue resulting in a large number of accidents.

Emphasis in regulatory policy has shifted over the years (Handberg, 1992.) Earlier (1976-83) health and safety regulation aimed primarily at improving the physical working conditions, i.e. establishing adequate washing and eating facilities, regulating temperature and noise levels, requiring the screening/shielding of machinery and the use of protective clothing. Initially, the inspectorates were also greatly involved in promoting the establishment of the joint management-labour health and safety committees in all relevant firms. Health and safety regulation during the last decade has focused more on improving the labourer-machine interaction, i.e. regulating work speed, work organization, planning and construction and defining work content. All these issues have traditionally been subject to negotiation in the collective agreements.

Health and safety regulation seems to be on a collision course with the private (wage-) agreements. the Federation of Danish Employers' Associations (DA) and the Confederation of Danish Labour Organizations (LO) have, in an attempt to reduce this potential conflict, agreed to implement an action plan/programme in connection with the agreements made on the labour market to reduce repetitive, monotonous work (DA & LO, 1992.) If the occupational hazards associated with working in the fish processing industry are to be significantly reduced by the development of clean technology, then the regulatory activities to promote these changes must also be negotiated with the parties on the labour market.

Case study design

Our sample consists of 16 fish processing industries, located in the county of North Jutland. In each case we have collected qualitative data from four different sources:

- plant inspections and open-ended interviews with management and with worker representatives;
- a review of all municipal documents regarding the firm's emissions, production permits, payment rules for waste water discharge fees;
- a review of all of the labour inspectorates' documents on these firms;
- a series of workshops for shop stewards and labour representatives from the firms' joint management-labour health and safety committees.

The regulatory history of each firm is based on a review of all *formal* correspondence with the firms, whereas the informal attempts to secure cooperative solutions to non-compliance, e.g. telephone consultations, have not necessarily been recorded. While these data give a good insight into when and what was required of the firms, the data from the workshops shed some light on the question of *why* firms have behaved as they have done in the regulatory processes and to what extent the employees have been involved in the clean technology development process.

The empirical results — uncoordinated regulatory action

An overview of legislation regarding health and safety as well as the environment revealed that the different authorities address some of the same problems. Each of these areas of regulatory overlap will be discussed in the following sections. There is of course a wider variety of problems which each of the two agencies seek to regulate and which have no bearing on the other agency's work. We will not comment further on these activities.

Project planning: The health and safety inspectorates require that new or renovated buildings satisfy certain legal and technical conditions. The inspectorates have primarily focused on the physical aspects, although they have in two instances required that the production processes — and subsequently the work processes — be organized according to certain specifications. The environmental authorities not only approve of the construction plans but also the planned emissions (air-borne, noise and solid wastes.) Firms are therefore required to monitor their emissions and report these to the authorities on a regular basis. The production/pollution permits, issued by the environmental authorities, are valid for eight years at a time, whereas the inspectorates continuously can require the firms to make specific changes if new information on the dangers of production becomes available. As a result, firms must comply with different types of regulatory demands with quite different time horizons.

The information which both of the regulatory agencies have at their disposal usually consists of architectural or technical drawings as well as details of the requirements laid down by other authorities such as the veterinary authorities. The environmental authorities and the health and safety inspectorates are supposed to notify each other of their respective regulatory demands, but our study indicates that the inspectorates are more informed about the environmental authorities' requirements than *vice versa.* The inspectorates received copies of the firms' pollution permit applications, and could — at least in principle — have required changes to be made, if they found the production conditions alarming. This has, however, not been the case there are no "correction notices" from the inspectorates in the municipal documents.

Air quality: The air in the production halls is renewed by ventilation and the in-take of clean air from outside. The working conditions have been improved by rapidly removing cooler air, smell, chemicals and humidity, while at the same time minimizing worker exposure to micro-organisms and aerosols. The environmental authorities have subsequently required firms to either build their chimneys higher or to install pollution abatement equipment. These requirements have not affected the working conditions or the health and safety regulation in any negative way, although improvements in the working conditions could have had a detrimental effect on local air quality, had the ventilation stacks not been built high enough.

Chemicals: Workers must be protected from the harmful short and long-term effects of chemicals. Accordingly, the firms have been required by the inspectorates to install ventilation and to issue protective masks to the workers. The environmental authorities have, on the other hand, concentrated their efforts on ensuring that the firms collect and dispose of their hazardous chemical wastes in an environmentally sound way. Both of the regulatory agencies have required the firms to store harmful chemicals in a safe manner, but have done so in a way that has not been in conflict with the other agency's requirements.

Noise: The tolerated noise level is much higher inside the plants than outside. The inspectorates have made requirements with respect to machine construction and/or the installation of noise abatement techniques. Workers have also been required to wear hearing protectors, whereas the environmental authorities have regulated working hours and required firms to comply with specific noise limits at plant boundaries.

In sum, the environmental authorities have primarily tried to reduce plant emissions and the use of water, whereas health and safety inspectorates have sought to reduce worker exposure. The regulatory agencies have had virtually no contact with one another when dealing with the above-mentioned problems, and have not coordinated their respective efforts to regulate firm behavior. During our interviews, the officers from labour inspectorates have made it quite clear that they have been unable to comment on the documents from the environmental authorities — let alone coordinate their regulatory activities with them. They have simply not had the time or personnel to take on this task. Other studies have indicated that this is quite typical (Handberg & Christensen, 1992).

It seems obvious that much could be gained with respect to time and expertise, if the regulatory agencies were to coordinate their activities in these areas of regulatory overlap. Presumably, it would be possible to simplify regulatory procedures, avoid juridical conflicts and perhaps even reduce the number of rules and regulations to be administered. Ideally, the government would be able to use its relatively limited resources more effectively. The agencies could for instance assist one another in monitoring firm performance, and on the basis of this common knowledge would hopefully be able to promote a more integrative approach to the development of clean technology than has hitherto been the case in the fish processing industry.

The firms and the development of clean technology

The local environmental authorities have played a rather modest direct role in stimulating the development of clean technology in the firms in our sample. Six of the 16 fish processing industries have developed and implemented clean technologies, primarily to reduce the amount of water used in the production processes and to separate the organic materials from the waste water. These innovative activities have been encouraged by consultants, and to a large extent, funded by government (state) subsidies through the Clean

Technology Development Programme 1986-89, administered by the National Agency for Environmental Protection.

The health and safety inspectorates have not been involved in these development processes at all, even though the environmental authorities notified them of these activities. This information is contained in the production/pollution permits issued by the environmental authorities and sent to the inspectorates. These did, however, not react to this new information and as a consequence, a variety of technological changes were made without any consideration being given as to how these would affect the working environment.

The development of clean technology depends not only the regulatory action of the two sets of authorities. It also depends on the agreements made by the parties on the labour market and the extent to which the employees (e.g. the firms' health and safety committees and cooperative committees) are involved in the development processes. Our case material from the workshops with shop stewards and health and safety officers indicates that there are a number of significant obstacles to be overcome before clean technology can be considered as a development strategy for the fish processing industry.

Firstly, work is based on an extensive sexual division of labour, and as a consequence, the most pervasive occupational hazards differ according to sex. The women suffer primarily from repetitive, monotonous work, poor work postures and work in cold conditions. Only half of them rotate jobs along the production line. The work speed is high. One-fifth of the women are piece-workers. Half of them are paid by the hour. The rest work at time rates with a bonus. The women have very little job security. According to the collective agreement, employers can send them home, if not enough fish have been landed. The women are not permanently employed, but are "casual workers", hired on a day to day basis. They have no influence on work organization. The men, on the other hand, have better working conditions (e.g.they rotate between a number of different job functions, the work rate is moderate and none of them receive piece-wages) and more influence on work organization.

The female employees suffer from occupational hazards which are closely linked to the way in which the work processes are organized. This is regulated through the agreements made between the parties on the labour market. This is not the case when it comes to the occupational hazards for men. These are closely related to the physical conditions in the production halls, i.e. associated with the noise level, changes in temperature and slippery floors, and are regulated under the Health and Safety Act.

When asked about what possibilities the employees saw for the development of clean technology, all agreed that this was of great importance, but that the clean technologies should be centered around improving their working conditions. Presumably the male workers will have a greater influence on these development processes than the women in the fish processing industry. If they are to participate and have any influence on this, then their working conditions will have to be changed. This will require labour market negotiations and compromises from both sides. This process of change is likely to be subject to inertia.

Secondly, initiatives to introduce clean technologies have been based on a top-down approach. Clean technologies were in the process of being introduced in six out of 13 firms, but the employees were only involved in two instances and in each case the decision to introduce clean technology was taken solely by management without any discussion with the labour representatives. The employees had tried to raise the issue in four other instances, but could not get their management's or employer's support. Asked about what obstacles might be in the way of an increase in worker involvement, they pointed to the following:

1. the attitude of management — underrating the importance of encouraging worker involvement in developing clean technology. Furthermore, the employment conditions makes it difficult for the workers to develop the necessary competence and obtain adequate training programs.
2. lack of awareness on the part of the workers — difficulties in changing their co-workers "forces of habit" has also helped to maintain the *status quo*.
3. lack of training — making it difficult for the workers to take part in the in-house discussions regarding the introduction of new technology.
4. the sexual division of labour — adding a gender dimension to all of the above mentioned problems.

There is very little in our case material to indicate that management or employers have taken the "integrative approach" and actively collaborating with the joint management-labour health and safety or cooperative committees. The latter are, according to the Agreement of Collaboration between DA and LO, responsible for discussing the implementation of new workplace technology. Representatives from either of these committees have if at all, been included at a relatively late stage in the development process. They have been confronted with almost completely developed "solutions" to the environmental problems at hand. Despite whatever possibilites the joint management-labour committees may create for labour representatives and management to air their differences, this has not been sufficient to ensure a more integrative clean technology strategy, that takes health and safety and environmental problems as a whole rather than separate problems to be solved one at a time. As a result, the firms will presumably have to use many resources to "de-bug" the clean technology solutions to the firms' environmental problems, once they are installed.

Other attempts to develop clean technology in Denmark have clearly illustrated that the introduction of clean technology is not just a matter of technique, and that dramatic improvements in the firms' environmental performance cannot be realized by relying entirely on the advice from outside experts. If clean technology is reduced to merely involving technical (process and/or product) changes, then it's potentials will not be fully realized. The success of the clean technology development strategy also depends on the "softer" elements/variables such as experience, knowledge and organizational structures. Thus, it may be necessary to change wage systems, work organization, training programmes etc. in order to gain access to the workers' experience and insight, to motivate them and utilize their creativity. All of the firms' internal resources have to be mobilized if pollution prevention is to be a success. This is, of course, hinged on the commitment of management (Cramer, Fischer & Schot, 1992).

Conclusion

Our exploratory study of the implementation of clean technology in the fish processing industry raises two important questions. First, in light of the environmental authorities and health and safety inspectorates' virtually non-existent efforts to coordinate their regulatory activities, how can government performance be improved? Secondly, what role can the firms play in ensuring the development of clean technology?

Prospects for government policy

The regulatory histories of the 16 fish processing industries indicate there is much to be done in coordinating regulatory action. Environmental control should take place in close collaboration with the work environment authorities. If this doesn't happen, then it will continue to be confusing and difficult for the firms to keep track of what demands are being made and by whom. Changes in environmental policy have, and will continue to accentuate this problem, simply because seeking to reduce pollution at its source will necessarily involve changes in the working environment. If regulatory action remains uncoordinated, firms will presumably continue to either optimize production from an environmental point of view or minimize worker exposure, but not do both at the same time. An obvious starting point for improving government performance could be coordinating plant inspections. This will, however, require one of two things: either giving the local authorities more resources or a political acceptance of changing the priority assigned to the different regulatory tasks.

Furthermore, our material indicates that two different regulatory "traditions" or styles of regulation have been in use, and that these differences in styles have also blocked a more integrative regulatory approach. As mentioned previously, the two sets of authorities have different ways of setting regulatory standards. Although both rely on "commands and controls," they do so in a different manner. The environmental authorities issue pollution permits. The health and safety inspectorates order specific conditions to be improved. The inspectorates are, however, not able to set these demands independently, but are to a large extent dependent on the parties on the labour market reaching an agreement as to what these demands might be.

The actors involved in the regulatory processes also differ. Environmental policy is more open to public participation than health and safety regulation and many different groups or organizations can influence the regulatory process, for example by complaining about the pollution permits issued by the environmental authorities and by appealing to higher levels of government. Health and safety regulation on the other hand, is limited to the parties on the labour market. A more integrated regulatory approach to the development of clean technology will also require the development of new political arenas within which the regulatory processes can take place.

Prospects for firm policy

The workshops have given us an insight into the problems associated with a clean technology development strategy, as the shop stewards and the worker representatives in the health and safety committees see it. Based on these observations, we conclude that the fish processing industry faces some substantial obstacles in developing a clean technology strategy. There are several reasons for this, i.e. the way employers/management manage health and safety at work, the management's and the work force's training and level of competence, as well as the structural constraints imposed on the individual firms through the agreements made by the parties on the labour market.

An obvious way of ensuring greater worker involvement would be to rely on the existing structures — the health and safety committees and the cooperative committees. This seems to be a very weak way of getting things done. Although the health and safety committees are supposed to be included in planning stages when changes in technology are being made and the cooperative committees are supposed to discuss how these changes will affect employment and training requirements, and relay this information to the workers, this has seldom been the case in the fish processing industry.

More drastic measures will have to be put to use. Something must for instance be done about the collective agreements. These have great influence on the employment policies pursued by firms (i.e. allow firms to rely on casual work), the training programmes to which employees are entitled, the possibilities of job enrichment and the work rate and thereby on the amount of repetitive, monotonous work. To overcome these obstacles and obtain better and cleaner jobs, it will be necessary to change the collective agreements, i.e. to add a sustainability dimension so as to ensure the appropriate changes in technology and work organization. "Clean collective agreements" could be one way of ensuring that firms adopt an innovative strategy in solving their environmental problems and improving working conditions, instead of short-sightedly treating these issues as something secondary to "doing business."

References

Arbejsministeriet og Miljøministeriet: Betænkning om virksomheders miljøtilsyn, Betænkning nr. 1193, [Ministry of Labour and the Environment: Report no. 1193 on Firm Compliance Monitoring] March 1990.

Handberg, S.: *Den nordjyske fiskeindustri — branchebeskrivelse*, [The fish processing industry in Northern Jutland — a sectorial description]TOR-projektet rapport nr. 1, Skriftserien Institut for Samfundsudvikling og Planlægning, Ålborg Universitetscenter, 1992.

Christensen, P., S. Handberg, L. Kromann & E. Holm Nielsen: *Miljø- og arbejdsmiljøreguleringen af den nordyske fiskeindustri* [Environmental and Health and Safety Regulation in the Fish Processing Industry], TOR-projektet rapport nr. 2, Skriftserien Institut for Samfundsudvikling og Planlægning, Ålborg Universitetscenter, 1992.

Christensen, P., L. Kromann & E. Holm Nielsen: 'Industrial Environmental Regulation and Innovation: A case study of the fish processing industry in Denmark', paper presented at the European Environmental Conference, Nottingham, 22-23rd of September, 1991.

Cramer, J., K. Fischer & J. Schot: *The Greening of Industry, Summary Report — First International Research Conference*, The Netherlands, 1992.

Handlingsplanen mellem DA og LO om det ensidigt gentagne arbejde, [Action plan for the reduction of repetitive, monotonous work], København, 1992.

Jacobsen, P.: *European Labour Law, Suppl. 33 — Denmark*, Kluwer, The Netherlands, 1984.

World Commission on Environment and Development; *Our Common Future*, 1987

6 The environment and business: The incorporation of environmental responsibility into retail ethics

Sally Eden

Introduction

Environmental issues were first taken up by mainstream business in the UK in the late 1980s (Marshall and Roberts 1992). Since then, environmental references have been included in product labelling, in product and corporate advertising and in company policy statements. A green image has become increasingly important to companies (Higham 1990) so that in a 1989 survey of businessmen, 40% of respondents believed that a green image was beneficial for their company and even "made commercial sense" (Hilton 1989). A contemporaneous survey reported that around 40% of companies claim to have environmental policies or objectives in place, 37% had undergone a complete or partial environmental audit and 29% had reported their environmental policies or performance in their Annual Report (Coopers & Lybrand Deloitte 1990 quoted in Carey 1992; Buck 1992) showing an increase in disclosure of environmental information since the early 1980s (Harte and Owen 1991).

However, the information disclosed in annual reports and other publications is rarely standardised and usually relates only to individual sites or instances of 'good' environmental behaviour. Owen (1992, p.20) describes this form of disclosure as "specific narrative" because, although potentially auditable, at present it reports only singular incidents not company-wide operations. The environment is now represented in very general terms, as in the following policy statement:

> "To conduct our affairs with real consideration for the environment, both in the products we sell and in all operational practices throughout the business." (Sainsbury's press release June 1991)

A key concept which is frequently advanced by businesses and retailers in such policy is environmental responsibility which merits further investigation as a possible basis for proenvironmental business change.

In this study, the environmental perspectives of retailers are investigated as well as the manner in which they use the concept of environmental responsibility. Data were collected from two types of retailers through in-depth qualitative interviews: national and local representatives of major superstore grocery chains having local stores in Leeds (hereafter multiple retailers) and small independent healthfood/wholefood shops in the Leeds area (hereafter small retailers). The data obtained from these interviews, and from material published by these and other grocery multiples, were analysed to reveal the major themes in environmental responsibility and proenvironmental change as perceived by retailers. (Further details of the methodology can be found in Eden 1992.) The data reveal how retailers relate business responsibility to consumer responsibility, to economics and to social expectations and how the environmental responsibility of business is therefore constructed. This chapter develops some of these themes.

Business responsibility

The responsibility of business was first defined publicly as a social responsibility for the company's effect on its employees and surrounding community (e.g. Sethi 1981), but is now increasingly also associated with environmental performance. Environmental responsibility is hence commonly slotted in by the multiples alongside other forms of responsibility because of the trend that this director perceives:

> "A successful organisation today demonstrates a responsibility in a whole spectrum of areas; environment clearly sits in amongst those areas."

With the exception of Safeway, whose business responsibility is "extremely selective" in prioritising the environment (Adams et al 1991 p137), this responsibility does not give the environment priority over other issues but includes it in decision-making. The environmental responsibility of business is clearly identified and stated by multiple retailers, but at a general level in their marketing logos, publications and the oral data in this study: it is nonetheless used specifically by ASDA in labelling products as "environmentally responsible". Small retailers discuss responsibility less explicitly, emphasising ethics which are stronger, more internally driven and less publicised than the responsibilities perceived by multiples.

The publicising of such responsibility has attracted much comment. Some commentators find business responsibility difficult to reconcile with the economic and profit-making priorities of business, and denounce responsibility as "fundamentally subversive" and inappropriate for business (Friedman 1988).

In the main, however, responsibility is seen by retailers and commentators as necessary (rather than morally correct) because of the power and position of business in society. This transforms (big) businesses from purely economic operations to social institutions, which must bridge the "legitimacy gap" (Sethi 1981) between their operations and the social expectations placed on them and thereby justify their existence and activities (Buck 1992, Smith

1990, Epstein 1981).

Sethi (1981) identified three different levels of business responsibility which promote legitimation by paralleling social expectations:

(i) a proscriptive responsibility to obey market forces of economics and legislation;
(ii) a prescriptive responsibility to operate congruently with the norms and expectations of society and to be one step ahead of these;
(iii) a proactive responsibility to "promote positive change" (*ibid.* p80), to be responsive to society.

Sethi's second category is essentially one of reaction and anticipation and incorporates the first category of compliance upon which prescription must build. Such reaction and anticipation is central to much of the retailers' perspectives. Smith (1990, p59) follows a similar line but combines the arguments of profit and responsibility by describing the extent of business responsibility in terms of the priority placed upon profit maximisation in relation to responsible or proactive behaviour.

Legitimation of business activities and responsibility is furthered by promoting acceptance of consumer sovereignty, in that advertising and marketing implicitly or explicitly use consumer sovereignty to justify actions which can be criticised on moral grounds (Smith 1990). This rests on the assumption of a one-way flow of instruction from consumers to other institutions such as retailers, manufacturers and government bodies (e.g. Adams et al 1991 p3, Heelas and Szerszynski 1991, Burke 1990). Here, the consumer is viewed as sovereign, as "the ultimate source of power in the economic system" (Galbraith 1972, p.218); Galbraith terms this the accepted sequence.

However, consumer sovereignty is limited by an individual's situation and constraints upon information and choice (Adams et al 1991, Heelas and Szerszynski 1991, Smith 1990 p295, Singer 1983). As business and other institutions play a large part in providing both information and choice related to consumption, their actions curtail consumer sovereignty (Singer 1983) because there is a two-way flow of instruction. Galbraith (1972) termed the flow of instruction from business to consumer "demand management" and described a revised sequence to assumed consumer sovereignty where business and other institutions promote social beliefs and individual goals which favour themselves, e.g. through promoting consumption of manufactured goods. This conceptualisation of producer sovereignty (Smith 1990 p34) is also taken up by Gorz (1988). Its critics thus argue that consumer sovereignty is flawed but still used by capitalist institutions and actors to legitimate their actions (Harte et al 1990, Smith 1990). In this, it is used to relocate the source of power from business to the consumer, therefore conceptualising business as purely reactive and amoral. Responsibility is relegated to a reactive one, downplaying the power of business.

The difficulties of teasing out consumer and business roles within a two-way flow of instruction using environmental references are illustrated by Adams (1992) who invokes both consumer sovereignty and business demand management in the same sentence:"The constant factor... is individual demand. This demand, though in part created by them, is the tune to which our major companies dance." (Adams 1992, p.107)

Business environmental responsibility may be assessed both by the extent of changes ascribed to it and the sources of responsibility in the context of consumer sovereignty and business legitimation.

The extent of retailer proenvironmental change

Although overall changes made by one multiple are perceived by its managers to have been "phenomenal", in general multiples have only undergone moderate changes to operations and products. The easy changes are those first adopted (also Buck 1992), for example, according to one director:

> "instead of taking the whole world and trying to cure it once... with a finite resource available, you have to determine priorities, so we'll do the easy bits first."

These changes are easy in that they involve little or no financial sacrifice. The pervading example is that of CFC removal from aerosols (and later from packaging and refrigeration systems), prompted by the Montreal Protocol signed in 1987, and by the development of chemical substitutes by manufacturers. These products were easily handled by the retailing system: public support and customer demand justified the change and made it easy in economic terms; legislation was perceived as an extra force necessitating such change for all companies. Similar reference can be made to the siting of public recycling banks in retail car parks, which are easy to administer and therefore require little financial expense to appear environmentally active. Voluntary changes to company cultures are likewise easy and may be the only initial evidence of proenvironmental change. Increasing awareness amongst management and the production of policy statements on environmental action for wider consumption by the public is now common practice for the retail multiples.

The notion that proenvironmental change can be easy (at least at first) is also conveyed to the customers where advertising and publicity for modified green products play down any sacrifice or change in performance, e.g.:

> 'It is easy to achieve a fresh, clean wash and care for the environment.' (ASDA publication advert)

> 'Changes [in products] so small you won't even notice them'. (Tesco Green Choice leaflet, January 1992)

As well as being easy, a second characteristic of multiples' change is that it is perceived to be gradual, even rather ponderous, and made up of a number of steps or minor changes. This has several implications. Firstly, it implies caution, due to the uncertainties of the green market and an emphasis on reaction not proaction and thus the adoption of only prescriptive responsibility. Secondly, it implies a longer term approach: companies are keen to publicise their record of (previously unacknowledged) environmental responsibility plus a commitment into the future. This seems in part a legitimation of the present level of moderate change, e.g. for these national level managers:

> "there will be the long term commitment of businesses like ours, who've grasped environmental responsibility with both hands and said, 'This has got to be part of the way we conduct our business.' Regardless of the economic climate, we will continue to down-size, down-gauge, reuse, reduce - all the things we have been doing."

> "This is not new for us... We certainly didn't turn onto environmental things, we've been doing some of these things for a long time as part of... our social responsibility."

A gradual process implies that the company intends building on changes and that the environmental policy is a permanent feature within decision-making although not necessarily a priority above other considerations.

Thirdly, environmental issues may be developing alongside other low-salience issues as part of an overall change in company culture, especially where major retailers are attempting to move upmarket, e.g. as one commentator wrote of Tesco:

> "The greening of Tesco is the result principally of research which showed that Tesco customers are becoming more sophisticated and critical." (Higham 1990, p17)

This refers to changes in the customer base or to the changing of attitudes of loyal customers but is likely to include elements of both as the company changes in response to wider public changes. As retailer change is only moderate and gradual, it is not particularly proactive in forcing ahead social change. Environmental responsibility thus only seems to be reactive, and linked closely to the benefits this accrues rather than justified ethically: this is worth examining.

Commentators such as Simms (1992), Blaza (1992) and Buck (1992) have advocated business proenvironmental change because it is positively good for business. The multiples likewise perceive that such change can be an investment in the future of the company:

> "The success of recent initiatives demonstrates forcefully that responsible environmental policies can be highly compatible with good business decisions and can enhance shareholder value". (Safeway Annual Report 1991, p29)

Short term costs, although probably readily identifiable and even measurable, are therefore accepted within a vista of long term benefits in social viability, as for this local store manager:

> "There are going to be certain things which are going to be costly and may not actually benefit the company in the short term, but I believe... looking long term it will be cost-effective, not only from a cost point of view [sic] but from a human point of view... it's got to be beneficial

to the company but not necessarily today, tomorrow."

Retailers may see the ultimate justification of proenvironmental activities and professed responsibilities in terms of sales benefits - the tangible, economic expression of approbation and legitimation. However, the data show that the retailers see economic spin-offs from proenvironmental change as more significant than product sales because minor changes can reduce inefficiencies or minimise waste and therefore reduce costs of production. For example, Safeway's energy conservation programme saved them about £1 million between 1988-1989 (Adams et al 1990) and for this national manager:

> "In commerce, by and large, investment in the environment is a money- earner... it's a very happy spin-off... in the sense that if we peak efficiency... then out of that comes minimised costs and therefore maximised profits".

This raises a problem for conceptualising the initiation of proenvironmental changes. The connection between the environmental benefits and the economic benefits of some actions is so tight that it becomes difficult to see whether environmental responsibility or economic incentive stimulated the change. In some cases, the perception of environmental responsibility initiates an environmental review which must then be rationalised economically. Whereas the environmental benefit alone would prove insubstantial and not initiate change, alongside cost incentives it can illuminate unseen potential. This does not presume the environment secures a higher priority than profit but the view for a national environmental manager is that:

> "it's an excellent spur to make us look again, very consciously, at a lot of these environmental issues and out of these things... shake out profits. You get rid of inefficiencies."

It seems likely that a priority operates both ways: in some cases environmental ideas illuminate a route towards cost savings, e.g. through increased efficiency, and in other cases cost incentives prompt actions which prove to have beneficial environmental consequences. In the latter case, it seems that the rising costs of landfill have driven forward reviews of in-store waste volume and handling which reduce some environmental impacts because:

> "by aggressively attacking recycling, there are opportunities to turn existing costs into revenue." (Marks and Spencer internal publication)

Multiples hence tend to labour the point that environmental and economic considerations can sit in easy balance, even that they complement one another in decision-making. A Safeway publication calls them "compatible", KwikSave calls environmental considerations "plain good sense" for business and this national director says:

> "Truthfully [change] has got as much to do with the economy as it's got to do with the environmental drive, because the two things really go hand in hand... whatever... is the most economic actually is the

most environmentally friendly".

The Government seems to support this as the recent White Paper on the environment states:

"There is... no contradiction in arguing both for economic growth and for environmental good sense." (Department of the Environment 1990, p.8)

This seems a rather simplistic view which other commentators have debunked, pointing out the "irreconcileable" (Higham 1990, p17) differences between environmentalist and business goals (Irvine 1989). This is noted in the data not by the multiples but by the small retailers who perceive themselves to be operating on the (illusory) compromise of selling the most environmentally and ethically sound products their economic situation allows. Thus, for this wholefood shop owner: "In society as we've got now, the only real way to run a green business is to lock your door and go home".

Not all the perceived benefits of change are measurable in terms of economic returns: less tangible benefits emerge through enhanced reputation, image, good publicity and customer loyalty. All the multiples perceive these to be part of an (indirect) investment in their company's long term economic viability. Perhaps more pessimistically, hazards of poor publicity are perceived if no change is undertaken and thus voluntary acknowledgement of environmental responsibility at least avoids such negative effects, e.g. for this national manager:

"You will not find, I'm sure, any major public company that does not make a strong environmental statement... if a large company does not make a statement, their shareholders and others will start becoming suspicious and raise questions".

In this sense, the declaration of responsibility may be a defensive measure to pre-empt criticism and earn legitimation, with the multiples perceiving the importance of taking "a more defensible attitude" so that:"It'd be difficult for people to point the finger at us and say we're doing all these nasty sort of things."

All these characteristics of change - easy, gradual, beneficial, defensive - point to a reactive responsibility in attempting to match social expectations but not forcing change ahead of these.

The sources of retail environmental responsibility

As well as change ascribed to environmental responsibility, it is necessary to consider whether this responsibility stems from internal or contextual sources and therefore whether proactive or reactive responsibility is in operation. In the data from the multiples, there is considerable emphasis on an internal source of environmental and social responsibility within the company culture so that: "the company thinks green, considers green... the company believes it's the right thing to do."

However, this is contradicted by the same multiples where they perceive their responsibility to be dependent upon definition by consumers (via their expectations) and subsequent approbation (via their purchases) so that the same manger assumes consumer sovereignty when he remarks that:

> "The source of change of anything is the consumer.. what that consumer wants, he or she will get."

Public expectations define what businesses must do to gain public acceptability on ethical grounds, so that even business's internalisation of such ethics is in essence a reaction to the social context society and not solely internal development. The obvious corollary to this is that when public expectations rise for individuals, they will also rise, although perhaps to a lesser height, for business. Business will have to change to become more morally acceptable, to live up to more exacting standards in order to gain public legitimation. (Of course, if expectations weaken with the rising priority of economic costs, standards of acceptability may be less stringent.) It seems most likely that business responsibility must be defined and evaluated both externally and internally through internal change and self-assessment, and not solely internally as some policy statements may imply. This twofold source is clear from this national manager: "We will do things which we think are morally and ethically correct, the sort of things which we think our customers will expect us to do."

The moral responsibility for environmental protection felt by retailers is therefore a reactive one, a response prescribed by the context of public expectation but necessitated by the retailers' need for legitimation in this context. This incorporates the obligation (the proscriptive responsibility, Sethi 1981) to obey economic and legislative principles, e.g. Sainsbury's is clearly stimulated by the Montreal Protocol on CFC production to state that:

> "Sainsbury's feels a very real sense of responsibility to do all it can to stop using CFCs in its business." (Sainsbury's publication 1989, also 1991)

Although positive, it represents movement in the direction of foreseen change and required compliance, rather than having a hand in establishing this direction. A common example is of the anticipatory move that retailers claim to have made in relation to CFC-free products and systems: but this was done whilst acknowledging the Montreal Protocol to reduce CFC production. Indeed, the CFC issue seems to have been the first major non-food environmental issue to prompt a response from the retailers (Simms 1992) and it is now accepted as standard and no longer a stimulus for change. Such anticipatory change is preemptive and offers a short term competitive advantage until the other companies catch up with the legislation; it is therefore good for short term profits and long term business reputation. This is also seen in relation to retailers' perceptions of forthcoming EC legislation on eco-labelling (now in draft and due for implementation on specified product groups by the end of 1992); compulsory environmental audits and statements (in draft according to Owen 1992) and restrictions on packaging (there is a draft EC directive to force the recovery of 90% and the recycling of

60% of packaging waste within ten years, *The Guardian* July 16 1992, p8).

In contrast to the reactive and anticipatory responsibilities outlined, a proactive responsibility to promote social change (Sethi 1981) is expressed by some retailers in their publications and oral data, both in terms of stocking new products and in modifications to in-store operations. Such changes are perceived by the multiples to be ahead of legislative requirements or market demand and the company sees itself as pioneering, particularly for Safeway: "we've acquired a growing reputation for being proactive, rather than reactive." (Safeway leaflet 1991)

The need for such proaction in the green market is also urged by business people, such as Blaza (1992) and Carey (1992), so that business in general can avoid legislative compulsion and that particular businesses can gain competitive advantage.

Critics of business and economic rationality (Galbraith 1972, Gorz 1988, Smith 1990, Wernick 1990) suggest that large companies are able to manage demand for specific products through information and advertising. This would represent proaction very clearly but there seems little explicit evidence in the data for this in the environmental field, except for Safeway's support of organic produce through stocking and research. There is an implicit suggestion that several multiples do attempt to influence demand but this is poorly developed. The multiples interviewed also emphasise the limits to any perceived proactivity as it applies only to certain issues or product sectors and cannot be pursued indefinitely without the back-up of demand. Although the small retailers also refrain from proaction, their desire to be proactive underlies their discussions more clearly than the multiples', for example:"So all we do is open a shop.. and that's it, that's as far as we go saying that you shouldn't do this [consume such damaging products]."

They may seek out alternative suppliers whereas multiples seem to expect more from their existing ones. More commonly, they promote certain products such as organics and provide channels for producers and growers to sell them:

> "We do keep trying to push organics... you're providing a service by being there, offering them an outlet and opening up the channels for growers to supply their products to people who want them".

Generally, they wish to convert customers, in a non-aggressive way, to vegetarianism or other ethical purchasing via promoting awareness of the environment and groups. This is especially true of one of the most recently established wholefood shops in the area, which explicitly aims: "to inform people as part of education... to promote these ideas and way of life."

For most multiples, a proactive stance is denounced outright as incautious and too far ahead of demand to be fulfilled. It seems that the multiples prefer to maintain a competitive edge, paradoxically, by following the proenvironmental trend of other companies but not risking out-and-out proactivity without the back-up of proven demand. Here, the aim is to keep up with the field, not necessarily to lead it, to be "conservative" in change.

Whether proactive or reactive, the fulfilment of environmental responsibility is limited by functional perspectives. The large retailers adopt a stance as choice- providers, seeing their role as one of providing alternative products and enabling choice not by dictating behaviour to customers but by:

> "offering them as an alternative to standard products to give customers a choice - 'environment friendliness' can involve a price premium, or some limitations in the product's function, and the Company does not wish to impose these restrictions by not offering an option." (Sainsbury's publication 1991, p5)

And for this national director:

> "It's not for us to dictate what our customers do and don't buy. It's not for us to say, 'We're not gonna sell you aerosols, because we think they're bad.' We're retailers. We have to satisfy the demand of the population we service, and they tell us eventually through their purchases, what they do and don't want."

The multiples seem to be abrogating their moral responsibility beyond these reactive levels. This suggests a perception of the retailing function as amoral because it should pass full responsibility to consumers. Similarly, Gorz points this out when discussing capitalism's incorporation of economic rationality:

> "It was no longer a question of good or evil but only of correct calculation. 'Economic science', insofar as it guided decision making and behaviour, relieved people of responsibility for their acts. They became 'servants of capital' in which economic rationality was embodied. They no longer had to accept responsibility for their own decisions." (Gorz, 1988, p.122).

This is shown in this comment from a national manager:

> "We are selling a [environmentally damaging] lifestyle... whether we like it or not, those are customer demands. Now, there's an argument about whether we were guilty of creating those demands and whether we just fulfil those demands and that's a moral argument."

So, the decision over what is morally acceptable is implicitly passed onto the consumer because of the need for legitimation (Smith 1990, Harte et al 1990), and hence survival, in the public domain. The multiples implicitly recognise their need for public legitimation and their abrogation of responsibility thus:

> "All sorts of things going on which we tell people about because it's all part of justifying what we do, otherwise we're always accused of being cynical and superficial and short term and only in it for the money... [but] it starts from within, it's part of the thinking, part of the culture and attitudes of the people in control of this business."

> "We don't sell ecologically based or environmentally based products in their own right as loss leaders to gain kudos or whatever. We make

things that we think are right... the things that are right for the customers, that the customers have been shown to demand. And if the customers don't like them, they don't buy them, we stop selling them."

It seems that ascription of responsibility may be more clearly made to business for its internal operation but actions in the public domain, e.g. product reformulation and marketing, are passed to the public as the ultimate repository of responsibility for moral legitimation of business responsibility.

Conclusions

Proenvironmental change for multiple retailers is usually moderate and gradual. The data indicates that, where environmental criteria match economic criteria, so that proenvironmental change yields positive benefits, environmental ethics are incorporated into retailers' ethics in the public domain. This legitimates the actions of business by contextual reference to social expectations, suggesting the adoption of a prescriptive responsibility (Sethi 1981). In this, the role preferred by multiples is one of reaction and anticipation: to customers, to legislation, to changes in the economy (Simms 1992, O'Riordan and Rayner 1991).

However, at the explicit level there remain contradictions between proactive responsibility, which is internal and often devoutly expressed, and reactive responsibility which is external and contextual. It seems likely that changes in a multiple's behaviour are neither strictly forced nor purely voluntary but are anticipatory, reactive to other changes and responsive ahead of their enforcement but recognising that such enforcement hovers in the near future.

References

R. Adams, (1992) Green Reporting and the Consumer Movement. pp..106-118 in D Owen (ed) *Green Reporting: Accountancy and the challenge of the Nineties*. Chapman & Hall. London

R, Adams, J. Carruthers and S. Hamil (1991) *Changing Corporate Values*. Kogan Page.London

A. Blaza, (1992) Environmental Reporting - a View from the CBI. pp.31- 34 in D. Owen (ed) *op cit*.

J. Buck, (1992) Green Awareness: an Opportunity for Business. pp.35-39 in D. Owen (ed) *Ibid*.

A. Carey, (1992) A Questioning Approach to the Environment. pp.87-97 in D. Owen (ed) *Ibid*

Corporate Intelligence on Retailing (1991) *The Retail Rankings*. The Corporate Intelligence Group. London

Department of the Environment (1990) *This Common Inheritance: Britain's Environmental Strategy*. HMSO London

S. Eden, (1992) *Individual Motives and Commercial Retailing in Green Consumerism*. Unpublished Ph.D. Thesis, Department of Geography, University of Leeds.

E. Epstein, (1981) 'Societal, Managerial, and Legal Perspectives on

Corporate Social Responsibility - Product and Process'. pp.81-103 in P. Sethi and C. Swanson (eds) *Private Enterprise and Public Purpose: an understanding of the role of business in a changing social system.* Wiley, New York.

M. Friedman, (1988) The Social Responsibility of Business is to Increase its Profits. pp.217-223 (reprinted from 1970) in T. Donaldson and P. Werhane (eds) *Ethical Issues in Business: A Philosophical Approach.* Prentice-Hall. New Jersey.

J. Galbraith, (1972) *The New Industrial State*. 2nd edition. André Deutsch. London

A. Gorz, (1988) *Critique of Economic Reason.* Verso, London.

G. Harte, and D. Owen (1991) *Environmental Disclosure in the Annual Reports of British Companies: A Research Note.* University of Leeds School of Business and Economic Studies Discussion Paper Series, No. AF91/04.

G. Harte, L. Lewis and D. Owen (1990) *Ethical Investment and the Corporate Reporting Function.* University of Leeds School of Business and Economic Studies Discussion Paper Series, No. AF90/07.

P. Heelas, and S. Bronislaw (1991) Buying the Right Stuff. *Town and Country Planning* 60 (7), 210-211.

N. Higham, (1990) *Marketing Week* 13 (19) July 20, 17.

A. Hilton, (1989) *Marketing* August 31.

S. Irvine, (1989) *Beyond Green Consumerism* Friends of the Earth Discussion Paper No. 1. Friends of the Earth, London.

T. Marshall, and P. Roberts (1992) Focus: Business and the Environment. *Planning, Practice and Research* 7 (2), 25-28

T. O'Riordan, and S. Rayner (1991) Risk Management for Global Environmental Change. *Global Environmental Change* 1 (4), 91-108.

Owen, D (1992) *The Emerging Green Agenda: A Role For Accounting.* University of Leeds School of Business and Economic Studies Discussion Paper Series, No. AF92/01.

P. Sethi, (1981) A Conceptual Framework for Environmental Analysis of Social Issues and Evaluation of Business Response Patterns. pp.69-80 in P. Sethi and C. Swanson (eds) *Private Enterprise and Public Purpose: an understanding of the role of business in a changing social system.* New York: Wiley.

C. Simms, (1992) Green Issues and Strategic Management in the Grocery Retail Sector. *Int. J. Retail and Distribution Management* 20 (1), 32-42.

P. Singer, (1983) Rights and the Market. pp.102-111 in V. Barry (ed) *Moral Issues in Business*, 2nd edition. Pub: publisher.

N. Smith, (1990) *Morality and the Market: Consumer Pressure for Corporate Accountability.* Routledge, London.

A. Wernick, (1991) *Promotional Culture: Advertising, Ideology and Symbolic Expression.* Sage, London.

7 Towards an environmental credit-rating agency

Joseph Tanega

Introduction

In the 1992 publication by Stephan Schmidheiny and the Business Council for Sustainable Development entitled Changing Course, the Chapter entitled, Capital Markets: Financing Sustainable Development, begins with a bowing acknowledgement to the mysteries of the capital markets and their relationship to environmental problems.

Capital markets will play an important role in the search for sustainable development, but little is known about the constraints, the possibilities, and the interrelationships between capital markets, the environment, and the needs of future generations.

Virtually no research has been done on this subject. It is essentially new territory, a fact that forces us to limit ourselves to some very basic arguments and conclusions. We thus offer an agenda for further research, analysis, and dialogue.[1]

There is over one trillion U.S. Dollars worth of financial assets traded daily in the form of money market certificates, treasury bills, commercial paper, stocks, euro-bonds and all other kinds of electronic chits circulating the planet.[2] This value constitutes a virtual world of trading information which is several magnitudes higher than the total global currency reserves of $550bn.[3]

Although Maurice Strong has stated that $380 Billion US per year is needed from the overdeveloped world to the developing world for global environmental protection, we have very little idea of where this money could come from. Taxes and government subsidies fall far short to cover this tab. There is not yet the sense of urgency to drive the political agenda to spend as much on environmental protection as on global defence (according to the United States of America General Accounting Office this was $980 Billion in 1990). To engineer a massive movement of funds into environmental protection will take a deep understanding of how the financial markets actually operate. Without delving into the mathematical financial models,[3] we shall take a practical anthropological approach and ask how are players in the capital markets responding to environmental issues; what are their systems of beliefs; and how can we devise financial products and instruments which will enable financial players to take environmental issues on board as part of their everyday business.

The primary premises of this chapter are:

(1) that the international financial and capital markets are engines and filters of a vast web of unequal information flows;
(2) that the main function of traders and fund managers (financial players, for short) in the capital markets is to create profits;
(3) that environmental information has heretofore been considered unconsciously by the financial players in the capital markets; and
(4) that new financial instruments (or equivalently, new information products) can now be devised to service the needs of the financial community for environmental information.

Before we proceed, we issue a warning that this paper does not aim to present the mathematical models of what these financial instruments may be.[4] The line to draw here is that we are arguing for the creation of new conventions in the financial and capital markets which may be initiated by regulation or by sheer market ingenuity.[5]

In either case, an elegant solution to the problem of funding environmental protection could be found by implanting the appropriate "social genetic algorithm"[6] which would automatically find the best solution in the marketplace. Taking the cue from Schmidheiny and the Business Council for Sustainable Development, we shall present an outline of how these new conventions may be initiated. Whether or not the markets will accept this issue is not of our concern.

We should not draw too harsh a judgment on the lack of hard data in the new field of environmental or ecological economics and if we were presumptuous enough to believe that we were approaching the union of the sciences in some mythical mathematically perfect Elysian fields, we ought not to forget that there is a huge gulf between the practice of international financial investments and economics itself. Academic economics and mathematical economics for the most part do not test itself against economic data.[7] It is a well known fact, that for all the use of supercomputers and mountains of statistics generated by government agencies, the theories of academic economists are utterly divorced from the grubby reality of daily commercial life.[8] To redress the balance, this chapter is also a call for more testing of regulatory and market-based solutions using the massive resources available for computer simulation.

For example, we note not without some delight how in 1982, Leontief surveyed four years of articles published in the most prestigious economic journal, The American Economic Review (from 1978 to 1981) and found over half the articles were mathematical models without any data and about one-fourth drew inferences from statistics gathered for some other purpose. Michael Rothschild comments, "Of the articles, only one-half of 1 percent--one article--used direct analysis of data gathered by the author.[9] That lonesome article was about utility maximizing behaviour in pigeons."

There may yet be a way for the human species to learn from the economising behaviour of other forms of life.[10] This chapter, however, does not propose any grand theory. In fact, the purpose of this chapter is to investigate a rather strange question. The question is: How can the individual trader in an investment house working with tradeable securities and financial instruments be connected to the concerns of global environmental protection?

If we can see how the concerns of the trader interface with global environmental protection then we might be able to see what programme of action we need to produce in order to effect, influence or accomplish our goal of making the capital markets more responsive to the emerging ethical values of environmental protection.

Let us step back a bit and view the capital markets from a wide-camera lens. What is its purpose and what does it do? In the simplest terms, the capital markets is a nexus between producers of goods and savers such that savers have an opportunity to invest to producers and producers have an opportunity to tap savers' funds. Fund managers and investment managers scan the virtual world of financial information for the most profitable investments. They are intermediaries. In ecological or biological energetics, we know that the carrying cost of processing information is very high relative to an organism's limited supply of energy. Specialised knowledge or processes which can more efficiently manage this overhead cost will be extremely valuable to the organism. By way of analogy, humans need to invest considerable resources to develop the expertise necessary to have a sense for the markets and to turn a profit. This is a highly paid skill. For example, the latest rocket scientists who are experts in 'financial engineering' and three to five years experience in swaps and derivatives may be paid $500,000 US per year basic with a bonus that is geared to the amount of profits generated by the small team. Excepting sports heros and media personalities, these young people are of the highest paid professionals per year of experience and practice in the world.

How does the trader in financial instruments take into account environmental protection considerations in his everyday decisions of buying and selling financial instruments? There are at least three different answers. One, the trader is not conscious of any environmental protection considerations since there are no means for him to obtain this information. Two, the trader who may have access to relevant environmental information is not concerned with using this information for his own business reasons. Three, the trader is using environmental considerations unconsciously since environmental considerations are already 'pre-figured' in the calculations of corporate value--the so-called 'market discount' theory.

Michael Silverstein, an environmental economist from the United States, argues for the third position. His argument is that since the United States has so many enforceable environmental laws and regulations, the cost of environmental protection is already discounted in a company's market price. For example, major chemical manufacturers, such as Dupont pay up to fifty percent of total pre-tax revenue into environmental compliance.[11] This is a viable way to measure the costs of environmental violations on corporate performance. Basically, we take the traditional concepts of contributions ratio analysis and account for the contribution of environmental costs as a ratio of total corporate revenue or profitability.

Indeed, the Investor Responsibility Research Center, Inc. (IRRC) of Washington D.C. have created a measure of corporate environmental performance for the Standard and Poors 500,(credit rating agency) in which the cost of environmental compliance is calculated for each of the corporations listed in the S&P. They have found that corporations which have higher than average compliance costs for their respective industrial sectors have either (1) old and inefficient environmental technology or (2) management which is not responsive to the environmental costs.[12] The likelihood of stock under-performance for these types of companies is statistically significant.[13]

The IRRC position is that there is a way to measure environmental compliance which reflects upon a corporation's overall profit or loss performance. Thus, we have the real life situation where instead of having a market with perfect information for everyone, there will be some people who know more than others about what is relevant for their bottom line. To put this proposition more forcefully, environmental information is itself a product within the capital markets system. It is a product which needs to be form-fitted for each of the major investment players in the capital markets field.[14]

For example, a fund manager who specializes in U.S. Treasures has no way of translating environmental information of the general state of the U.S. into marketable information which is relevant to his trading activities. However, on the other end of the spectrum, a specialist in "tradeable emissions" (or "pollution permits") will deal directly in the valuation of a financial product which is directly relevant to the quantity and quality of particular pollutants. Between these two end points is a whole spectrum of companies with equity and equity derivative products which are affected by the levels of compliance (risks of liability, costs of replacement, costs of new to environmental laws, fines and clean-up) to environmental laws.[15]

The original question of how the individual trader is connected to global environmental protection, therefore, narrows to a question of what types of instruments could be devised for traders to become more aware of the problems of global environmental protection. Obviously, this particular information product will need to have a direct bearing on the trader's methods of evaluating a company's, sector's or index's profit performance. This product will also have a direct bearing on a diverse field of users who are looking for a convenient and cost-effective way to access environmental information which is relevant to their investment portfolios.[16]

Eco-financial instruments

There are at least two levels in which eco-financial products may be developed for the financial and capital markets:

(1) As a fund management tool, a type of instruction set which guides investment managers in their daily task of stock picking; or
(2) An industry wide instrument which enables all financial players to immediately recognize the environmental value of a company.

In respect of 1, we have a multiplicity of funds which are ear-marked for environmental "plays." For example, a so-called "pure environmental play" in the parlance of investment managers is a company which delivers environmental clean-up or protection technology. Thus, a company specialising in soil remediation would be considered a "pure environmental play." In fact, it would be quite plausible to draw up a list of waste management, water clean-up, soil remediation technology providers, and invest in the best valued of their ilk and offer to sell this information or create an investment company based upon this concept. Call it "The Pure Environmental Trust Fund."[17] At the moment, the current state of the art for pure environmental plays is to find small niche companies supplying environmental technology which is sought in other jurisdictions. Thus, the environmental funds are helping the transfer of environmental technology throughout the world.

With regard to 2 above, the task of defining a viable environmental credit-rating agency is theoretically interesting since it calls for an understanding of how an "eco-credit rating" could be used by financial and capital market players. Since no such rating exists, let us begin in the normative and describe what types of information this rating should incorporate. Some may object at this juncture and say that a normative rating system would not be acceptable to the large players in the capital markets. However, in terms of market intelligence, the acid test is not sheer size or market dominance, but reliability and accuracy of market intelligence regarding perceived needs of customers.

The eco-financial rating system

Successful plays in the capital markets are not irrational and are more effective in processing relevant information about the markets than other people. For all the economists who still hang on to the Efficient Market Hypothesis and the Rational Man Standard, we can only mark Nietzsche's words of warning and announce once again, these gods are dead. These explanatory premises do not lead to cost-effective predictions.[18]

One of the most effective predictive (self-fulfilling oracles) functions in the capital markets is the credit rating agency. There are apocryphal stories about how the government officials of New York City are not so much afraid of crime or corruption as the power of Moody's (credit rating agency) to downgrade the City's debt. With a fall in rating, the down-rated entity will experience a higher borrowing cost.

Notice the emphasis is on logico-temporal necessity and there are no "if's" and "maybe's" as in most areas of life.[19] The risk of lower and lower downgrading of credit, the borrowing costs could escalate such that the entity would not have sufficient cash flow to meet future obligations on debt service, thus driving it into an unvirtuous spiralling cycle of higher debt service and eventual bankruptcy. We note that Moody's and Standard and Poor, the two largest credit rating agencies have on their current books at any one give time ratings of over 50,000 corporations, public entities, including government debts. The size of this portfolio represents the value of over ninety percent of all tradeable financial products throughout the world, and thus, represents far and away the very life-blood of total capital and trade flows for the whole planet.

The mechanics of the environmental credit rating service will obviously have to take into account environmental laws and regulations, their actual and potential enforceability, the corporations' past, present and future environmental commitments and programmes,[20] the situs of liability and so on. From these inputs, a rather mechanical "risk management" model could be derived and the weighted risk would be set against particular episodes in the future of the company. One can easily imagine the cocktail mix of proven technologies: an adaptive learning database, hypermedia satellite links and automated feedback loops in strategic corporate locations. However, without falling deeper into this information architectural "folie", suppose we were to devise a scheme for rating corporations and public entities on their environmental policies and implementation, find a valuation for their authentic commitment and compliance, and somehow make it relevant to the considerations of ethical investment managers. Suppose further that this rating system gained credibility throughout the financial and capital markets. What effects, if any, would this environmental rating system have on environmental protection?

First, the environmental rating system would allow investors to "punish" environmentally unsound companies and "reward" environmental sound companies. This mechanism therefore would have a type of "negative feedback" on the way investment management is now currently practised, adding an element of measurable corporate investment responsibility in respect of environmental issues.

Second, this rating agency would have the power (very similar to Moody's and Standard and Poor's) of immediately influencing the direction of investments without regulatory enforcement costs or additional taxes. In the United States, the number of environmental regulations which have cascaded from the Federal, through the State and to the municipal level has reached over 80,000 pages since 1981.[21] Do we need more regulations or more monitoring of existing regulations? The US Environmental Protection Agency is reported to employ over 35,000 workers of which 16,000 are lawyers. Moody's employs less than 1,500 and rates over 20,000 corporations.

Third, the unimpeachability of credit rating agencies is a measure of their strength as standing on the highest ground in terms of credible information flows. They provide a service of condensing huge quantities of information into chunks of information which the individual traders on the trading floor can respond to. Thus, vast quantities of information is reduced to decisions at the individual level, which means that individual investors who rely on investment managers will have more say in the direction of their investments. The environmental credit rating agency would be a step in giving investors an opportunity to give value to and enable the growth of more socially responsible corporate organisations, making them more accountable to global environmental protection.

Fourth, creative fund managers will have the opportunity to design new Eco-financial Indexes analogous to the S&P 500 Futures and Options which will allow traders to trade blocks of environmentally sound companies, thus increasing the momentum of capital on behalf of environmental protection.

Conclusion

In this short chapter, we have outlined the possibility and some of the benefits of an environmental credit-rating agency. If one considers the magnitude and costs of global environmental protection and the need for individual investors to have an economic "vote" in the capital markets system, then it is high time to begin constructing financial instruments which will meet these needs. I think it was Heisenberg (of the Heisenberg Principle of Uncertainty) who said, upon hearing that an atomic bomb was being built, that it was a lot of rubbish. However, that night he worked out the theoretical mathematics and in the morning announced, "Ah the atom bomb--it's just an industrial problem now." In many ways, the construction of an environmental credit-rating agency is now and always will be just an industrial problem.

We also note that if one percent of total global capital flows were allocated to environmental protection, then we would have more than enough "funds" to support all environmental protection projects within two weeks! With this system in place, there would be no need for new massive taxes or subsidies.

Ultimately, global environmental protection will be won by individual efforts. As we sleep our money circulates round the Earth unconsciously traded by individuals who have no established conventions within their professions which will allow them to support the environment. The creation of an environmental credit-rating agency would be the first steps towards a conscious evolutionary ecological economics. When the Environmental Credit-rating Agency speaks, all will listen...

Footnotes

1 Schmidheiny, S 1992. *Changing Course A Global Business Perspective on Development and the Environment*, The MIT Press, Cambridge, Massachusetts, p. 54.

2 Note here that the actual total value of transactions recorded in electronic media has not to my knowledge ever been calculated with any precision. The totality of value represents in some analogous way the totality of useable economic "energy" in the global economy. I estimate $1.5Trillion U.S. on the basis of summing the total value of transactions on the capital markets, the value of tradeable goods and the value of individual cash transactions interfaced with banking or credit institutions. Any and all valued transactions entered in the electronic media are parts of an electronic ecological economic system, with positive and negative feedback loops of information. The development, growth and evolution of financial instruments is premised on the ideal that rules of replication and natural selection apply to this field of information flows. Some instruments will have "genetic" capabilities and many others will not. The introduction of an environmental credit rating agency will have the effect of curbing transactions in line with global environmental protection values.

3 There are literally thousands of articles and books in this area. As an investment practitioner, I have much sympathy for the position of treating the macroeconomic environment as a dynamic system which means that literally all main macroeconomic variables are simultaneously determined within a structure in which the role of policy and exogenous variables are explicit. For a dynamic systems model which has been tested on the hard edge of the financial markets, see Karakitsos, E(1992) *Macrosystems: The Dynamics of Economic Policy,* Blackwell; Oxford.

4 The parameters of these instruments are however dependent on the risk of litigation and liability per jurisdiction, the short- and long-term effect of voluntary corporate announcements on publicly quoted share price, regulatory diligence of enforcement and so on. In theory, the probabilized risk of liability and effects of announcements are mapped against daily risks of cash flow, which in turn can be mapped against maturity dates of say debt instruments. For the effects of information disclosure by top 100 US corporations, see Lev, B *"Information Disclosure Strategy,"* in California Management Review, (Summer 1992) pp. 9-29.

5 Given the fact that the value of derivatives instruments has grown to over $20 trillion U.S. within the last five years with very little regulation, one wonders whether the regulators are intellectually capable of understanding what is actually going on. Consider the case of the 19 year old hacker who is able to breach the security codes of banks, defence departments and so on from his extremely low level technology

platform. Now imagine the same individual armed with finance theory and playing with the assets of banks. All banks and portfolio managers managing global portfolios are likely to have some form of quantitative modelling within the next five years. The people making crucial decisions of buying and selling securities are less than 30 years old.

6 For genetic algorithms see Holland, J H. *"Genetic Algorithms"* in *Scientific American*, (July 1992) pp. 44-50. In brief, genetic algorithms have been developed for the capital markets in order to gain greater and greater accuracy in predicting prices. By analogy, a social genetic algorithm is a set of rules which allow groups of people, who by randomly trading information, to find the most appropriate answers to specific problems. Setting an award or a prize serves only to set the goal; the social genetic algorithm is more aligned to setting the shape of trading information flows. An environmental credit-rating agency will change the shape of information flows such that financial information will be explicitly influenced by the rating generated from the environmental credit-rating agency.

7 Rothschild, M L. (1992) *Bionomics, The Inevitability of Capitalism*, Futura Publications: London, p. 52.

8 *Ibid*, p. 53.

9 *Ibid*, p. 53.

10 A systems view of how other forms of life have applied economising behaviour by creating ecological niches and how humans may profit from understanding complex systems design, see Mollison B (1988) *Permaculture: A Designers' Manual*, Tagari Publications: Tyalgum, Australia.

11 Silverstein, M 1992. Private communication.

12 Naimon, J, 1992. Private communications. See IRRC, Suite 600, 1755 Massachusetts Avenue, N.W., Washington, D.C. 10036.

13 *Ibid.*

14 This view is echoed by the announcement of the Hundred Group of Finance Directors of the U.K. one week after the UNCED Rio Summit June 1992, calling for minimum standard standards of environmental reporting in the annual reports of publicly held companies.

15 The Council of Economic Priorities, New York, is now publishing environmental risk profiles of the companies in the S&P 500. Each company is assessed in terms of its Toxic Release Inventory schedules and compared against companies within the same industrial sector.

16 *Ibid.* The main purpose of the environmental compliance ratings published by the Council on Economic Priorities is to inform institutional investors of the environmental records of companies which are within the investment portfolios of major institutional investors. This institutional investors may use such environmental risk reports as the basis for putting pressure on the companies to change their management's commitment to environmental issues.

17 I have investigated about 35 environmental funds. For those interested,

one might check the sales brochures and literature of: Global Environment Fund, L.P., Washington, D.C.;Kemper Environmental Services Fund, Chicago, Illinois; Oppenheimer Global Environment Fund, Denver, Colorado; Kinder, Lydenberg, Domini & Co., Inc., Cambridge, Massachusetts; Calvert Fund, Bethesda, Maryland; Jupiter Tarbutt Merlin, London, UK; Oko-Invest, Lindengasse 43/17, A-1070 Wien.

18 Casti, J J. (1991) *Searching for Certainty, What Science Can Know about the Future*, Scribners: London. See especially, Chapter 4: "Meanwhile, Over at the Casino," pp. 192-252, which is a very good layman's summary of the current state of financial engineering in the capital markets.

19 For the view that credit rating agencies merely confirm the general knowledge of the market see Wakeman, M L. (1986) "The Real Function of Bond Rating Agencies" in *The Revolution in Corporate Finance* ed. by Stern JM and . Chew,D H Jr. Basil Blackwell: Oxford, pp. 17-20.

20 Note how the proposed European Commission on Eco-Management and Audit Regulation, although voluntary, will require applicant companies to ensure that all contractors to conform to the equivalent standards of environmental management practice (Section 13 of Annex I). This rule will have a snowball effect and strongly encourage other companies to report their environmental performance as well. For detailed analysis of the effects of the Eco-audit Regulation and BS 7750 Environmental Management Systems on corporate reporting policies, see the author's work on Introduction to Environmental Management Systems published by Kingston Business School, London.

21 Smart, B (1992) *Beyond Compliance*, WRI: Washington D.C.

Part III

EUROPEAN ENVIRONMENTAL POLICY: A COMPARATIVE ASSESSMENT

8 Environmental policy in transition: A comparative overview of key developments in Britain and Germany

Matthew Gandy

Introduction

This paper examines contemporary environmental policy in Europe with particular reference to energy, water and waste sectors in Britain and Germany. I argue that despite significant differences in the policy making context in Germany and the UK there are two interrelated developments common to both countries: firstly, the emergence of political demands for higher environmental standards, resulting in increasingly stringent legislative standards; and secondly, mounting concern over the growing cost of environmental protection, in a context of wider attempts to limit public expenditure since the 1970s. A key outcome of these developments for environmental policy has been a modification of the role of the state in relation to the private sector, with an emerging process of 'de-municipalisation' in local government, as the operational aspects of environmental policy are increasingly handled by the private sector, and at the same time, an extension of the role of the state at a national and EC level in order to provide a more effective regulatory framework for environmental protection.

The UK experience

In the post-war years there was an extensive programme of nationalisation in the energy, water and waste sectors in response to the perceived inability of the private sector to provide a range of essential services adequately and undertake the capital investment needed in order to modernise and improve standards. Yet by the mid-1970s the political consensus over Britain's public utilities began to ebb. There was mounting disquiet over the structure and function of the public utilities in terms of their responsiveness to policy changes, their accountability to the public, and their sapping of the public purse. The background to the massive and on-going programme of privatisation in the 1980s can be traced to two important developments in the mid to late 1970s. The first of these was the concern over the inability of the public sector to adequately fund the capital intensive water, energy and environmental protection sectors of the economy without disrupting wider attempts to control public expenditure and limiting resources for other services such as health and education. The second key development was the rapid growth of the environmental movement during the 1980s, as illustrated in Table 1 and political demands for higher environmental standards, both from grassroots public concern and also at an international level through pressure of the EC, the UN and international conferences and protocols. The outcome of these twin pressures has been an extensive privatisation programme effectively placing the burden of capital investment into the private sector coupled with the emergence of a new regulatory framework to oversee the privatised utilities and respond to the environmental aspirations of the public. The UK privatisation programme has extended to a wide range of key sectors integral to any coherent environmental policy strategy: oil, gas, electricity, water, and waste management (with the tendering for the collection and disposal of waste). Yet whilst the role of local government and the operational aspects of the public sector is in decline, the role of government is being simultaneously modified and extended at a national and EC level in order to impose higher environmental standards.

Environmental Group	**Individual Membership**		**% Change**
	1980	**1989**	**1980-1989**
Greenpeace	10,000	320,000	+3,100
Friends of the Earth	12,000	120,000	+900
World Wildlife Fund	51,000	202,000	+296
Ramblers Association	36,000	73,000	+103
National Trust	950,000	1,750,000	+84
CPRE	27,000	44,500	+65
RSNC	140,000	205,000	+46
RSPB	321,000	433,000	+35
Total	1,547,000	3,147,500	+103

Table 1 The growth of environmental groups in the UK during the 1980s
Source: McCormick (1991)

The declining direct involvement of the public sector, and particularly of local government, seems set to continue in the 1990s. The Government's proposed National Environment Agency, combining Her Majesty's Inspectorate of Pollution, the National Rivers Authority and the local government Waste Regulation Authorities will see a further reduction in the role of local government in the regulatory aspects of environmental policy. In the case of waste management it appears that the removal of the local authority role in the regulation of solid waste management is based on the pretext that local government has neither the resources nor the expertise to carry out uniform environmental regulation to create a 'level playing field' for business activity (Department of the Environment, 1991). A key question for the newly emerging nationally-based regulatory framework is whether it will be adequately resourced in order to implement the Government's environmental aims as set out in the 1990 White Paper and comply with the ever growing range of environmental directives issued by the European Community.

A further concern over the regulatory framework for the privatised utilities is whether the price rises for electricity, gas and water can be adequately controlled. In the case of electricity, there is doubt whether the structure of the electricity supply duopoly can allow sufficient competition to keep prices as low as possible for the consumer and also to allow a significant expansion of renewable sources of electricity and the achievement of wider environmental aims. Some commentators have suggested that a further fragmentation and decentralisation of the electricity supply industry into regional units serving local communities might assist in the promotion of combined heat and power (CHP) schemes for example, as in Sweden (Armstrong and Cooper, 1990). Certainly this theme of a community role in demand management for energy use through elected local authorities is not likely to emerge given the decreasing role of local government in these aspects of public policy. Concerns have also been expressed by the gas industry regulatory body, OFGAS, suggesting that there might be a regional structure created to facilitate competition, but this is predicted to reduce the value of British Gas shares and would therefore be fiercely resisted by shareholders (Beavis, 1993; Roberts et al, 1991). Indeed the lifting of any possibility of re-nationalisation or tighter regulation under a Labour government following the Conservative election victory of 1992 saw a dramatic leap in the value of shares in a range of privatised utilities, as shown in Table 2. An important question is whether there is now a conflict of interest developing between the short term interests of shareholders in the privatised utilities and the wider interests of the community, as reflected in a tougher regulatory regime for both consumer prices and environmental standards.

Utility	**% Rise**
Anglian Water	+ 80
London Electricity	+ 50
Manweb	+ 73
National Power	+ 19
North West Water	+ 78
Northumbrian Water	+ 76
Norweb	+ 72
PowerGen	+ 20
Yorkshire Electricity	+ 77

Table 2 Rise in share prices of selected privatised utilities (10 April 1992).
Source: The Financial Times, 11 April 1992

An important aspect of electricity privatisation has been an attempt to develop different sources of energy. In particular, the Non Fossil Fuel Obligation (NFFO) levy marks an example of strategic state intervention, which whilst primarily used to promote nuclear energy has also accelerated a number of developments in other non-fossil fuel sectors of the energy industry, as shown in Table 3. Note the proliferation of energy from waste schemes, principally from the harnessing of landfill gas and the production of electricity from waste incineration. These developments represent not only profitable alternative sources of energy, but have also been identified by the Watt Committee on Energy as a key component in any strategy to limit greenhouse gas emissions (Watt Committee on Energy, 1990). However, energy from waste schemes have been widely criticised by radical environmentalists for their failure to tackle the problem of waste at source in the production process.

Form of Energy	Number	Capacity (%)
Hydro power	26	11
Landfill gas and bio-gas	33	36
Waste incineration with energy recovery	7	46
Wind power	9	7

Table 3 Projects supported by the renewables price levy
Source: Office of Electricity Regulation (1990)

The Conservative election victory of 1992 has also resulted in the abolition of the Department of Energy and the splitting of its functions between the Department of Trade and Industry and the DoE. This may have important implications for the formulation and effective achievement of a national energy policy, since a powerful and unified government department is considered to have more leverage than a fragmentary administrative structure in policy implementation (Roberts et al, 1991). An issue of concern is whether sufficient emphasis will be given to energy conservation and the demand side of energy use, since this is widely regarded as both a cost-effective and integral element in any strategy to reduce greenhouse gas emissions and levels of acid precipitation and other pollutants. Indeed the scope for greater energy efficiency has long been recognised as fundamental to any attempts to alter the relationship between the use of energy and other resources and levels of economic activity and future prosperity as part of practical attempts to achieve the goal of sustainable development.

In summary, the UK is undergoing a transformation in the ownership and structure of a range of essential services, involving a growing operational role for the private sector and a modification of the role of the state to regulate the activities of private companies in order to achieve environmental and other ostensibly non economic policy objectives. The long term consequences of these changes may have profound implications for the capacity of government to implement environmental policies which conflict with the short term interests of the private sector.

The German experience

In the 1970s successive German governments passed laws in a number of areas, including the implementation of new legislative frameworks for the management of water and solid wastes. During the 1980s and early 1990s the Kohl administration has steadily expanded the German legislative framework for the environment and has responded to the environmentalist challenge to established policy making by creating a new Ministry for Environmental Protection and Nuclear Safety. In contrast to the UK, the German proportional voting system for parliamentary elections has allowed the representation of the German Green Party in the Bundestag since 1983, thereby enabling direct environmentalist pressure on the Kohl administration during the 1980s (see Mintzel and Oberreuter, 1990).

In the case of the energy supply industry, the structure and ownership can be contrasted with the privatised duopoly and regional electricity boards in the UK. In Germany there are numerous energy supply utilities functioning as private companies but having a complex set of relationships with government at central, regional and local levels through both regulation and share ownership. In the city of Hamburg, for example, the city has a 100 % shareholding in the local electricity supply company, Hamburger Electricitätswerke (HEW), thereby enjoying considerable leverage over energy policy for the citizens of Hamburg (see Gandy, 1992). In the field of air pollution, the German power station clean-up programme will amount to DM 21 million over the period 1983 to 1993, and most of this cost has been borne by electricity consumers facing an extra 1.3 Pf/kWh on their electricity bills. The planned UK programme for the retrofitting of flue-gas desulphurisation and low nitrogen oxide burners would have amounted to some £1.6 billion, equivalent to less than a quarter of the sum spent in Germany. Yet even this is being steadily scaled down through the use of low sulphur imported coal and the 'dash for gas' with the installation of combined cycle gas turbine power stations. In Germany, however, the installation of extensive FGD has made the burning of coal and lignite more environmentally acceptable and may serve to prolong the life of some of the existing power stations rather than encourage the expansion of combined cycle gas turbine plants.

In the 1970s the legislative framework for water standards was greatly strengthened in West Germany, and a number of improvements have been noted: by 1990 some 92% of the West German population were connected to waste water treatment plants and the waste water of about 86% of households was being biologically treated; the eutrophication of surface water has been reduced; and there has been a reduction in the level of heavy metals in rivers, particularly Cadmium, and the oxygen content has risen (Simonis, 1991). In other areas, however, there has been a deterioration in standards. An example is ground water which has been widely affected by pollution from nitrates.

Furthermore, the reunification of Germany has posed a huge challenge for the achievement of uniform standards across Germany:

> The problem for the East German water industry today is simple but acute - how to square traditionally heavy demand for water with a supply which is both limited and desperately polluted. The only realistic solution will involve huge long-term subventions from Bonn. (Ballay, 1992 p.11)

The best available data source on water quality in the former DDR, the Ost-Berliner Instituts für Wasserwirtschaft, found that in 1989 the water supply to over half of the country's 17 million citizens failed to meet the GDR quality standards, which did not even extend to cyanide, mercury or pesticide residues. Some 42% of water courses and 24% of lakes in the five new states are thought to be unusable as raw water sources, as a result of out-moded production methods and defective wastewater treatment facilities. German reunification has not brought any immediate improvements, indeed two different standards for water quality are laid down within the unification treaty itself: the authorities in Brussels and Bonn accept that east German water cannot meet the 1980 EC standards, and have allowed exemptions for 12 limit values, including arsenic, lead and mercury until 1995 (Ballay, 1992). It is estimated that over DM 80 million is needed to bring drinking water and wastewater treatment standards up to those of the western Germany. Another DM 6,000 million is required for the upgrading of the main drainage system, and it is thought that over 800 new sewage treatment works need to be renovated or rebuilt for towns and cities, along with some 3,000 redeveloped and improved works in the rest of the country (Opie and Plenderleith, 1991). This and other components of the cost of German reunification have however led to higher German interest rates in order to control inflation generated by the large increases in public expenditure, and this has contributed to instability in the foreign exchange markets and indirectly served to weaken the progress towards European Monetary Union and the ratification of the Maastricht Treaty.

Following the passage of the 1972 Waste Management Act, and later amendments stressing the need for improved waste treatment, increased recycling, and greater waste avoidance, there have been a number of identifiable developments in the waste management field: in 1970 there were some 50,000 uncontrolled landfill sites but by 1990 this figure had fallen to 500 closely monitored landfill sites; the number of special treatment plants and landfill sites for toxic and hazardous wastes has grown from 10 in the early 1970s to over 100 in the 1980s; the number of municipal incineration plants has doubled from 24 to 48 and now burn around 30% of the total municipal waste stream; and the level of recycling for municipal waste has increased steadily, with particular advances for glass, tin plate, oil and dry-cell batteries (Simonis, 1991).

These developments must be placed alongside a number of continuing difficulties for German waste management. Despite the construction of new facilities for the handling of toxic wastes, Germany has emerged as the world's leading exporter of wastes, with over half a million tonnes exported to some 50 countries in 1991 as part of a growing trend of 'waste colonialism' from developed to developing countries. The emergence of 'waste colonialism' can be attributed to three main factors: the growing disparities in environmental standards (and costs of environmental protection) between the West and elsewhere; the increasing liberalisation of world trade and facilitation of waste transfers; and finally, the desperate need of hard currency on the part of less developed countries and the former states of the Soviet Union and eastern Europe (see Bernstorff, 1989; Vidal, 1992). There has also been widespread alarm over the potential threat to health from dioxin emissions from incineration plants, and this has served as a further stimulus to attempts to reduce the size of the waste stream (Spill and Wingert, 1990). Despite the increased level of recycling, the municipal waste stream has increased from 16 million tonnes in 1970 to over 34 million tonnes in 1990, leading to charges that recycling does not in any way challenge the underlying dynamic of waste production at source (Moser, 1990). The particular problem posed by the growing proportion of packaging derived waste has led to the new German packaging law which came into force in June 1991. The so-called Töpfer law lays down stringent requirements on industry to take back and recycle packaging but initial evidence suggests that this has simply fed into higher prices for consumers; there may be serious conflicts with EC competition policy stemming from restrictions on the import of foreign goods; and there are already signs that the secondary materials markets are being swamped by materials, thereby undermining the economic viability of existing recycling programmes within Europe (*The Economist*, 30 November 1991; Hibbert, 1992).

These legislative developments in Germany illustrate that the limits to waste management policy in local government may lead to new forms of state intervention at a national level in order to respond to political demands for higher environmental standards in waste management (though this trend may be undermined by movement towards greater subidiarity within the EC). Indeed, the principle of recycling is being extended under new legislation to consumer durables, particularly cars and white electrical goods, in order to force companies to take back and recycle as much as is technically feasible. It remains to be seen whether the recycling of packaging will affect the growing contribution of incineration and private sector solutions to the problem of packaging waste. The present legislative framework and political context for recycling and waste management in the EC appears to be in a state of flux: whilst there is a process of 'de-municipalisation' affecting the operational aspects of waste management in local government, there are new forms of state intervention emerging at a national and international level in response to the perceived inadequacy of attempts by the private sector to control the waste stream. Whereas the 1980s was marked by a degree of consensus over the rapid expansion of materials recycling using the 'bring' system of on-street collection facilities, the 1990s is seeing a divergence between different conceptions of appropriate forms of recycling and waste management.

Conclusions

In the UK, the regulatory role of the state is drifting from local to national government as a consequence of the perceived inadequacy of local government to implement the new national and EC level environmental standards and ensure a competitive and uniformly regulated environment for the private sector companies who have the expertise and resources to pursue high technology options and implement complex 'end of pipe' environmental control procedures. In Germany, the local and regional state would appear to be more capable of responding to the environmental challenges of the 1990s, yet there is pressure to release more of the operational aspects of service provision to the private sector, whilst retaining the regulatory structures of local and regional government relatively intact. In this sense it could be argued that the 'de-municipalisation' of environmental policy has advanced further in the UK than in Germany, yet the underlying pressures to control expenditure and raise environmental standards can be identified in both cases.

One important theme to emerge from this paper is that the increasingly stringent environmental standards in the core economies are contributing towards a relocation of the burden of environmental externalities such as pollution emissions and toxic wastes outside the US, the EC and Japan to developing countries. This suggests that research which is focused exclusively on developed economies may misinterpret the environmental consequences of policies at a global level. A further area for consideration is the need to examine the changing role of the state in environmental policy, and the impact of wider political and economic developments such as global economic recession and the moves towards closer economic integration in Europe on the evolving regulatory and legislative framework for the protection of the European environment.

References

Ahrens A, Gonsch V, Ohde J, Vierth I, Viole J, Vödisch A and Zöllner G 1989 *Müllkonzept.* Billstedter Umweltnasen; Bund für Umwelt- and Naturschutz (BUND) Hamburg und umweltschutzgruppe Phyzik/geowissenschaften, Hamburg.

Akkre E, 1991 *Thermal recovery - part of an integrated waste management strategy.* Alliance for Beverage Cartons and the Environment, London.

Armstrong N and Cooper M 1990 Energy for the 1990s: A New Policy Framework. *Local Economic Policy Review*, Loughborough.

Ballay U 1992 Double Standards. In *The Water Bulletin* 514: 10-11.

Beavis S 1993 Break-up sought for British Gas. In *The Guardian*, 2 March 1993.

Bernstoff A 1989 *Der neue Abfallkolonialismus - Müllexporte der Industriestaaten.* Paper presented to the conference Fachkongress:

ökologische Abfallwirtschaft. Technische Universität Berlin, 30 November - 3 December 1989.
Boehmer-Christiansen S and Skea J 1991 *Acid Politics: Environmental and Energy Policies in Britain and Germany.*
DoE (Department of the Environment) 1990 *This Common Inheritance: Britain's Environmental Strategy.* HMSO, London.
DoE (Department of the Environment) 1991 *Improving Environmental Quality: The Government's proposals for a New Independent Environment Agency.* HMSO, London.
Drösser C 1990 Der Wohlstand müllt sich tot. In *Geo* 7: 40-62.
EC (European Commission) 1989 *A Community Strategy for Waste Management.* SEC(89)934. European Commission, Brussels.
ENDS 1992 Brussels makes concessions in new draft on packaging. In *The ENDS Report* 204, January 1992.
Friedrich H 1989 *Die Müllverbrennung - eine ökologisch verantwortbare Technologie?* Paper presented to the conference Fachkongress: ökologische Abfallwirtschaft. Technische Universität Berlin, 30 November - 2 December 1989.
Freie und Hansestadt Hamburg 1989 *Abfallwirtschaftsplan.* Baubehörde, Amt für Entsorgungsplanung, Hamburg.
Gandy M 1992 The recycling of household waste: urban environmental policy in London and Hamburg. Unpublished PhD thesis, London School of Economics.
Goodhart D 1991 Germany moves in front. In *The Financial Times*, 14 August 1991.
Hibbeln K 1992 The development of waste management in Hamburg. Paper presented to the Second International Forum on Resource Recovery from Waste. Imola, 21-24 September 1992.
Kinnersley D 1988 *Troubled Water.* Hilary Shipman, London.
Leipert C and Simonis U 1990 Environmental Damage: Environmental Expenditures. Wissenschaftszentrum fur Sozialforschung, Berlin.
Mintzel A and Oberreuter H 1990 *Parteien in der Bundesrepublik Deutschland.* Bundeszentrale für politische Bildung, Bonn.
Moser M 1990 Environmental relief through waste prevention. Unpublished paper. Institut für ökologisches Recycling, Berlin.
Mukhopadhyay A K 1975 The Politics of London Water. In *The London Journal* 1: 207-226.
Newham M and Knight S 1991 A double dose of German recycling. In *The Financial Times*, 30 January 1991.
Opie R and Plenderleith I 1991 Eastern promise. In *World Water and Environmental Engineer*. December: 13-15.
Rasmussen M 1991 *Waste to energy technology.* Alliance for Beverage Cartons and the Environment, London.
Roberts J, Elliot D, and Houghton T 1991 *Privatising Electricity: The Politics of Power.* Belhaven Press, London and New York.
Simonis U E 1991 *Environmental Policy in the Federal Republic of Germany.*

Wissenschaftszentrum, Berlin.
Spill E and Wingert E. (eds) 1990 Brennpunkt Müll. Sternbuch, Hamburg.
Theil W 1988 Kosten/leistungsaspekte und geeignete Einsatzgebiete der getrennten Naßmüllsammlung und Kompostierung am Beispiel Hamburgs. Unpublished thesis, University of Hamburg.
Thornhill J 1991 German packaging law condemned. In *The Financial Times*, 28 June 1991.
Valette J 1989 *The International Trade in Wastes: A Greenpeace Inventory*. Greenpeace, Washington DC.
Vidal J 1992 The new waste colonialists. In *The Guardian*, 14 February 1992.
Watt Committee on Energy 1990 Renewable Energy Sources. *Report No.22*. Elsevier Science Publishers, London.
Wittern I 1990 Der Modellversuch Komposttonne (Grüne Naßmülltonne) in Hamburg - Harburg. Paper presented to the conference *Fachkongress für Energie und Abfall*. Neumunster, Germany, 7-8 June 1990.

9 Global warming and energy policy - Burden or opportunity?

Ute Collier

Introduction

The signing of the climate convention at the UNCED in 1992 binds countries to formulating and implementing strategies to mitigate climate change. A number of industrialised countries, including some member states of the European Community, already began elaborating response strategies previous to the convention thus allowing a first analysis of their scope and potential effectiveness.

A major emphasis in the response strategies has to be on energy policy due to the large contribution of energy activities to emissions of greenhouse gases. There is much scope for emission reductions from these activities, mainly through fuel switching and improvements in efficiency. However, there are currently a range of obstacles preventing such changes, such as the failure of the market to take environmental costs into account. Response strategies thus need to address these obstacles through regulatory, fiscal and other measures. Traditionalists in the energy sector may see this as a burden, while others see it as an opportunity to bring about a more efficient energy system.

This chapter looks at global warming response strategies in Germany, the Netherlands and the UK. It highlights policy conflicts between measures necessary to reduce greenhouse gas emissions and other objectives in energy policy, focussing on those affecting the electricity sector. The chapter then aims to identify options available to meet both environmental and other objectives, leading to better policy integration. Conclusions are drawn about the likely future direction of energy policy.

Energy activities and greenhouse gas emissions

Globally, energy production and use are estimated to account for over 50% of greenhouse gas emissions (Fisher, 1990), mainly through CO2 emissions from fossil fuel combustion. Accordingly, governments often attach priority to emission reductions from this area. The transport and electricity sector are the fastest growing energy sectors and in neither there is a quick technological fix to reduce greenhouse gas emissions. In the electricity sector there is a vast array of possibilities to reduce emissions on both supply and end-use sides but few offer large potentials on their own. Hence, a strategy incorporating a whole range of options is required. The global warming issue certainly provides a challenge but at the same time also an opportunity to develop a more efficient and sustainable energy system.

Traditionally, planning in the electricity sector has been orientated towards supply extension. However, most power plants operate at efficiencies of only 35-40%, with most of the primary energy wasted in the form of low-grade heat. Here emission reductions can be achieved either through improving the efficiency of fossil fuelled plants (eg. through advanced coal technologies, use of efficient gas turbines, more use of combined heat and power (CHP)) or switching to other generating options such as nuclear (which in itself remains controversial), or renewable technologies.

Efficiency on the end-use side had in most countries improved considerably as a response to the oil price shocks of the 1970s but efforts slowed down during the 1980s. Now, concern about global warming seems to be giving a new impetus to end-use efficiency and in the European Community (EC) there is much interest in the demand-side programmes pioneered by US regulators and utilities. At present there are large consumption differentials of up to 50% between the existing stock and best available technology for many electrical appliances like refrigerators (IEA, 1987). For certain applications (eg. heating) technical potentials of up to 90% savings are suggested (Enquete-Kommission, 1990).

At present these options face a number of obstacles. A general problem is that emissions and their consequences are treated by the economic system as so-called 'external costs' and are hence not reflected in energy prices. Thus, environmentally beneficial generating options are only chosen if they offer other benefits like economic viability. Assessments about efficiency are generally only made on narrow economic grounds (Collier, 1993). On the end-use side, there are also a number of other problems such as lack of information, divergent decision making criteria between consumers and suppliers, split responsibilities, limited access to capital and so on (Fisher, 1990; Jochem and Gruber, 1990).

Governments generally have recognised that greenhouse gas emission reductions cannot be achieved by relying on the operation of the market alone and that global warming response strategies have to involve intervention in the market through a variety of measures such as subsidies, provision of information, adequate energy pricing (through energy/carbon taxes and appropriate tariffs) and regulation (eg. technical standards). Governments seem to have become increasingly aware that environmental problems like global warming cannot be dealt with on an *ad hoc* basis but need comprehensive policy approaches and integration with other policy areas (see CEC, 1990).

Yet, energy policy continues to primarily pursue other objectives. Recently, there has been a shift away from concerns about security of supply towards moves to introduce competition in the traditionally monopolistic energy sector. This approach seems be in direct disagreement with the need to intervene more in the market to deal with greenhouse gas emissions. The following sections examine global warming response strategies in three European Community countries and investigate whether they are conflicting or compatible with other energy policy agendas.

Responses to global warming in the Netherlands

Concerns about global warming developed in the Netherlands at a time when environmental issues were already high on the political agenda and the government was in the process of drawing up its first comprehensive National Environmental Policy Plan (the NEPP). The possibility of sea-level rises has of course particular relevance to the Netherlands and it is therefore not surprising that the issue was taken up immediately by politicians.

The NEPP set a target for the stabilisation of CO_2 emissions at 1989/90 levels by the year 2000 at the latest. Additionally, the NEPP stated that ultimately, emissions of CO_2 in the industrialised nations must be reduced by 90%. With this, the Netherlands were one of the first countries in the world to publish an officially endorsed CO_2 target. However, the NEPP did not elaborate on any new instruments to achieve this goal.

Following the elections in 1990, the target was amended in a revised NEPP to the following:

- Stabilisation of CO_2 emission levels in 1994/95 at their 1989/90 levels (182 million tonnes).
- An absolute reduction in CO_2 emissions in the year 2000 to the level of 173-177 million tonnes p.a., that is an effective reduction of 3-5% (VROM, 1990).

These objectives are calculated against the background of a projected rise in CO_2 emissions of approximately 2% annually, linked to a 2.5% annual rate of economic growth. Three-quarters of the reduction is expected to come from the energy sector through fuel switching and energy efficiency, while 15% and 10% of the reduction respectively are to be achieved by the transport and the waste sectors (VROM, 1990). In terms of measures to achieve these targets, priority is to be given to the most cost-effective measures and those which have also other positive effects on the environment or serve other social needs.

As in other countries, the Government decided that energy efficiency would have to play a major role. The Government thus approved a memorandum on energy conservation outlining how this is to be achieved. In general, little time has been wasted in implementing the CO_2 reduction strategy, consolidating existing measures with a number of new measures.

The first new measure which was introduced in 1990 was a small carbon tax of DFl 5.70 per ton CO_2 and Dfl. 044 per GJ levied on fossil fuels. This tax raises DFL 150 million each year and is used to help finance the rest of the

CO_2 strategy (VROM, 1992). Secondly, a whole range of financial incentives are available such as subsidies for end-use efficiency measures (eg. insulation, condensing boilers, Combined Heat and Power (CHP) renewables and beneficial technologies. Thirdly, the Government runs energy training courses for important actors like small firms and construction firms. Additionally, it has reached an agreement with the electricity sector on emission reductions (see below). Overall, DFL 524 million were available for the strategy in 1991. Further measures are under investigation, including efficiency standards and the possibility of linking energy efficiency to environmental licences which are required by factories.

The Netherlands has a tradition of voluntary agreements between government and industry which is again being applied in the case of global warming. As part of the agreement, energy companies have drawn up their own environmental plans for achieving reduction targets on CO_2 and other emissions. The government has authorised distribution companies to levy a small surcharge on gas and electricity prices to partially finance the environmental plans.

The measures implemented to promote energy efficiency objectives also meet the energy policy objective of security of supply. However, the global warming concern also had a negative impact on this objective. Until recently, the government had intended to increase the role of coal in electricity generation in order to reduce the dependence on gas but this has now been abandoned, essentially because it would have involved an increase in CO_2 emissions. In other areas of energy policy, however, environmental issues still seem to play no role. Collier (1993) gives as one example the recent restructuring of the energy sector where the government failed to take account of environmental concerns and thus possibly missed the opportunity of creating a structure better suited to environmental objectives. Nevertheless, the restructuring has had some positive effects, mainly by giving distribution companies the responsibility for small-scale electricity generation, including that from renewables and CHP. Nevertheless, these effects are largely incidental.

Responses to global warming in Germany

The first step taken by the German government was to establish an enquiry commission in 1987 into the future of nuclear power in the early 1980s. The Commission for the Protection of the Earth's Atmosphere (Enquete-Kommission "Vorsorge zum Schutz der Erdatmosphëre" - EK) recommended a 80% reduction in the 1987 level of CO_2 emissions by 2050, with 30% of this to be achieved by 2005 (Enquete Kommission, 1990).

On the basis of the EK's final report, the German Cabinet decided on 13 June 1990 to substantially decrease energy-related CO_2 emissions. An inter-ministerial working group (IMA) chaired by the environment minister was established and given the remit to elaborate a strategy for reducing CO_2 emissions by 25% until 2005. IMA published its first report in October 1990 and this, together with the changed situation due to German unification led to a revised cabinet decision (Kabinettbeschluss) on 7 November 1991. The targets set were a 25% reduction goal for the old Länder (states) and a 'substantially higher' percentage reduction for the new Länder, leading to 25-30% overall (BMU, 1990).

Government policy on CO_2 was again reiterated in January 1992 in a parliamentary paper in response to the latest IMA report. This stresses once more the need for implementing economic instruments with emphasis on subsidy programmes and CO_2 emission charges. Here the Government broadly supports the EC energy/carbon tax proposals (BMU, 1992).So far, action on CO_2 has failed to move beyond government pronouncements towards the actual implementation of measures. It is unclear at the moment exactly what measures will be adopted. One problem is certainly the strain on financial and administrative resources from German unification which is likely to limit the scope for measures unless additional revenue is made available through, for example, the application of a carbon tax.

Some measures which will be beneficial in greenhouse terms are already underway but these were initiated for other reasons. The Government currently spends 300 million DM on renewables research and development. Included in this is a subsidy programme for wind energy aiming at 250 MW of installed capacity by 1996. However, subsidies for energy efficiency and CHP have all been cut at federal level (apart from some assistance available for the new Länder) and information campaigns alone are unlikely to yield the necessary result.

Meanwhile, the energy sector (in particular the eight large regional energy companies), already feeling under threat from EC liberalisation proposals, is opposing any substantial measures which would affect its operation. Oppostion is particularly directed against the possibility of a carbon tax. However, there are many examples of efforts by small local energy companies to implement so-called local energy concepts which aim at an energy system that meets environmental concerns (for examples see Collier, 1993). The basis of these concepts is a mixture of demand-side management, CHP plants and renewables. However, such concepts may be threatened as they often depend on cross-subsidisation within the vertically integrated local utilities, which the Commission of the EC is trying to ban.

Responses to global warming in the UK

When concerns about global warming were first voiced in 1988/89 , the Government did little more than refer to the role of British scientists in developing the scientific understanding of the issue. The House of Commons Energy Committee published a report on the energy policy implications of the greenhouse effect in July 1989 which criticised the Government's policies especially the low priority given to energy efficiency and renewables (House of Commons Energy Committee, 1989) but its recommendations were not taken up by the Government.

Calls for action, including targets, continued but these were rejected until Margaret Thatcher announced, quite unexpectedly, in May 1990 that the UK was to stabilise its CO_2 emissions by 2005. The target was criticised as being too weak, and for making action conditional on other countries being prepared to take similar action. In spite of the low priority accorded to global warming in the election campaign of spring 1992, Major's re-elected government finally announced on the 30th of April that it would bring forward the stabilisation target to 2000. At the same time it hinted that it would place a levy on gas and electricity bills to pay for energy efficiency.

The Environment White Paper of 1990 was intended to spell out the Government's strategy to deal with global warming. In the event it failed to do so. While it contains a section on global warming, it did little more than reiterate the CO_2 reduction target and list existing measures. The paper mentions two main policy objectives which are to press for international action to reduce greenhouse gas emissions through an international framework convention, and to promote measures which encourage energy efficiency (HM Government, 1990). However, there are no targets for specific fuels or sectors against which progress can be measured.

The White Paper claimed that the Government was taking 'major initiatives' on energy efficiency. However, of the three countries studied, energy efficiency has been most neglected in the UK. There are only very limited financial incentives for businesses and low-income households available from the Energy Efficiency Office. Additionally, there is a programme to reduce the total energy bill for the government estate by 15%.

Additionally, the White Paper stated the aim to increase the contribution of renewables to a capacity of 1000 MW by the year 2000. Renewables currently for the first time receive government support beyond the research and development stage under the so-called non-fossil fuel obligation (NFFO) which was established to protect nuclear energy in the privatised electricity sector. The NFFO has a number of shortcomings which are described by Elliott (1992). An important point is that it has not been designed with environmental considerations in mind.

The government has meanwhile published annual reviews of the White Paper but these have offered little in the way of new measures. The latest statement on the global warming issue is a discussion paper on climate change (published by the Department of the Environment in December 1992) with the stated intention to stimulate public debate before the elaboration of the response strategy to be drawn up under the climate convention. This paper lists cost-effectiveness and minimum government intervention as the guiding principles of government policy, with an emphasis on voluntary action and economic instruments rather than regulation (Department of the Environment, 1992). This reflects general government philosophy on environmental and many other issues but at least there is recognition that voluntary action with no change in energy prices is unlikely to improve energy efficiency. In any case, apart from running a publicity campaign, the Government still has not moved any closer to actually implementing a global warming strategy.

The growing concern about global warming coincided with the privatisation and restructuring of the UK electricity sector. As in the Netherlands, the concern hardly featured in the deliberations, with mixed results. Increased competition has encouraged a switch to gas-fired generation, with much new investment in combined cycle gas turbines (CCGTs), which is beneficial in CO_2 terms. However, at the same time, there is a clear disincentive for companies to invest in end-use efficiency measures as their profits are directly related to the quantity of electricity sold. Also companies have little encouragement to invest in renewables or CHP.

Towards better integration?

Although there are many differences in approach between the three countries, some common themes can be detected. For one thing, end-use energy efficiency has been almost universally recognised by policy-makers as needing to be the cornerstone of global warming strategies. Currently, most planning in the electricity sector of the three countries is concerned with the supply-side and most utilities are sceptical about the possibilities of demand-side management (DSM). This reluctance by utilities to become involved in DSM may be due as much to unfavourable regulatory framework conditions as to general inertia,a feature of companies that enjoy a monopoly. It is clear that such problems need to be addressed through policy action, but as yet there is little evidence that this is happening.

Apart from greater end-use efficiency, other potential options for greenhouse gas reductions considered in the three countries include an increased role for low emission generation technologies like nuclear, renewables or combined heat and power. These again face serious obstacles. In some cases the economics are prohibitive but even where technologies are viable, obstacles exist. For example, large monopoly utilities are often not interested in decentralised, small-scale technologies. In the case of nuclear energy, the major obstacle is public opposition (although in the UK the main problem has been the high economic costs) and there are serious concerns about the overall environmental effects of this technology. Greenhouse gas emission abatement is thus very complex as there are no quick, easy solutions.

As Collier (1993) has shown, global warming is not the only challenge currently facing the energy sector in the three countries being discussed. The sector simultaneously has to contend with moves by individual governments and the EC Commission to liberalise its operation and to introduce energy companies to a more competitive regime. Both agenda challenge the *status quo* of the energy sector and their realisation would entail major changes in operation and structure. Liberalisation calls for an end to the monopoly position which most utilities in the EC enjoy while global warming calls for a reorientation of the utilities' operation away from supply-side planning toward integrated demand and supply-side planning.

The two parallel agenda are not necessarily compatible. Liberalisation, in its extreme form, coupled with deregulation, relies on the operation of the market to achieve an optimal operation of the electricity sector. However, as there are numerous other market failures apart from a lack of competition at present, the result would not necessarily be beneficial in global warming terms. If global warming is to be seriously pursued as a policy objective, the extent of liberalisation would have to be limited and a strict regulatory and/or incentive framework would have to be established, possibly similar to that in the United States. Examples like the US or the UK demonstrate that profit maximisation in a privately owned electricity sector has environmental and societal side effects which need to be checked (Collier, 1993).

Despite emerging conflicts at the policy-making stage, neither the global warming concern nor the trend towards liberalisation have so far had actual, far-reaching effects for the electricity sector in the EC (with the exception of privatisation in the UK). In both cases, the agendas are still very much under development and the eventual outcome is uncertain. Meanwhile, the sector continues to be expected to meet other, more traditional policy aims, in particular security of supply. For the future, at least four different outcomes can be envisaged, depending whether the three separate agendas are integrated, as can be seen in figure 1.

Figure 1: Possible future directions for the EC electricity sector

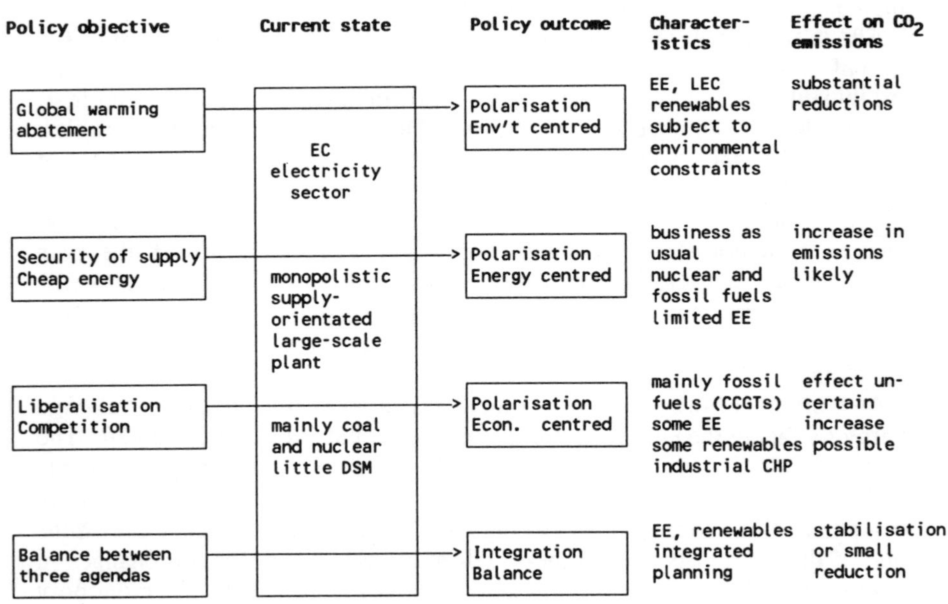

If any one of the three agenda dominates policy-making, policy polarisation is set to continue between environment, energy and other economic objectives. However, while one likely outcome may be a traditionally energy centred approach, another possibility would be for environmental objectives to take precedence. If liberalisation objectives were to dominate, narrow economic considerations would come before both energy or environment objectives. However, a better integration between the different policy agendas can also be envisaged.

The four paths would be likely to differ in terms of technology, fuel choice and structure of the electricity sector. Differences would still exist between countries but a number of general observations can be made.

The environment centred path would concentrate on end-use energy efficiency (EE), renewables and small-scale, efficient technologies, including CHP. Market forces might play some role but only where they could be exploited to the benefit of the environment. The electricity sector would possibly become more decentralised along the lines of the local energy concept (LEC) idea, with horizontal as well as vertical integration and planning. Lifestyles may have to adjust considerably to reduce wastage and hence energy consumption. This would then result in substantial CO_2 emission reductions.

With the energy centred approach decision-making would be dominated by issues like security of supply and low energy prices. In some countries, especially those without indigenous fuels and where public opposition can be overcome, this might include the nuclear option. Again energy efficiency and renewables might be promoted as these can contribute to meeting security of supply objectives. However, these two options would be less important than in the environment-centred approach. As this scenario would mean little change from the present situation, CO_2 emissions from the electricity sector would continue to increase.

If liberalisation was to become the overriding concern, separate demand and supply side planning would also continue as especially the generation side would become very diffuse. Energy efficiency and renewables would play some role but they would only be pursued if they could be shown to be economically viable . The result on CO_2 emissions is unclear as a reduction would not be specifically planned for. It is even possible that emissions could increase.

Finally, there is the possibility of a more balanced, integrated approach where policy objectives of all three agenda would be balanced, obviously with a number of compromises. This would result in a mix of the other three paths, ie. of energy efficiency, renewables but also some of the 'harder' technologies of the second path. Planning would become more integrated and might be pursued through an approach similar to least-cost planning as practised in the US. Some degree of liberalisation would be included, especially to encourage independent generators and DSM companies. This approach should at least stabilise emissions but reductions would be less than in the first path.

Although there would be differences in the level of energy efficiency and renewables, these two options nevertheless fit all three policy agenda. They are the least environmentally damaging options, can contribute to security of supply and additionally are suitable to a more liberalised regime as they offer opportunities for new companies to enter the market (as operators of renewable plant or as DSM companies). They are thus ideal options for so-called 'no regrets' policies and integrated policies aiming at the minimisation of conflicts. However, considering the many obstacles currently preventing the wider application of these obstacles, governments will have to intervene with instruments such as regulatory or fiscal measures.

The analysis of global warming response strategies indicates that EC countries are moving towards decision-making which contains all three elements and is therefore becoming better integrated. Examples of this process are the joint energy and environment minister councils at EC level or cabinet committees in the UK. However, in three cases, the Netherlands, the UK and at EC level, there has been a clear failure to integrate global warming concerns with attempts to liberalise or restructure the electricity sector. It emerges clearly from the analysis that global warming concerns are far from being a central objective in energy policy and more priority is attached to either traditional energy concerns or liberalisation. Too much reliance is placed on incidental benefits of measures taken to further these objectives. Generally, measures to reduce greenhouse gas emissions are contemplated if they happen to further other energy or general economic policy objectives, such as security of supply, diversity or cost-effectiveness. Basically, policy approaches are currently still not balanced with other policy objectives taking precedence over environmental ones. The outcome is a mixed picture with some positive effects in terms of GHG emissions. However, at the same time there are negative implications which could have been avoided if they had been addressed initially.

Integration, while easily adopted in principle, is far from easy to implement in reality. It is probably easier at the policy-making level, although this could result simply in lip service being paid to environmental concerns. At the implementation stage, ie. the move from general principles to the implementation of specific measures, policies are likely to fail if economic costs are perceived as being too high. Industrial interest groups can in all countries exert considerable influence during the policy-making process and often only have short-term private interests at heart. Furthermore, there are also practical problems of how to achieve integration between the various institutions involved in the policy process such as government ministries and regulatory bodies.

Conclusions

The analysis of the response strategies in the three countries reveals that at present, a real commitment to greenhouse gas abatement is lacking. The issue is seen by many decision-makers as a burden on energy policy and the energy industries, in particular in terms of a threat to cheap energy which is seen as essential for a country's international competitiveness. Proposed measures are thus subject to vigorous lobbying by industrial interest groups who threaten with relocation and loss of jobs. This results in adjustments only being made at the margin. The above discussion has indicated a number of areas of conflict between global warming response strategies and other policy objectives.

The main conclusion of this chapter is that a fundamental problem is presented by the fact that policy measures to deal with global warming are expected to work within a framework that has been created with other priorities in mind, and is generally based on narrow perceptions of economic efficiency. As there are a few complementary developments, the generally unambitious targets for emission reductions may be achieved under these circumstances. However, it is argued that any more substantial reductions, which are not to compromise other environmental concerns, will require a reorientation of the dominant supply-oriented, fossil fuel and nuclear based energy regime towards a real commitment to energy efficiency and renewable energy sources. For these reasons, the energy industry generally sees the global warming concern as a burden and as a threat to its status quo. However, governments seem to be increasingly questioning the status of the energy sector anyway and even without the global warming issue, changes are in stall, or have even already been implemented as in the case of the UK. Rather than a burden, the different agenda may be seen as an opportunity for change towards a more sustainable energy future but only if environmental concerns are given equal priority to other objectives. At present such an outcome is still very uncertain and the opportunity may well be missed, with potentially serious consequences for the environment.

References

BMU (1992) *Verminderung der energiebedingten CO_2-Emissionen in der Bundesrepublik Deutschland.* BMU, Bonn.

CEC (1990) Communication from the Commission to the Council on energy and the environment. *COM (89) 369 final*, 8.2.90.

Collier, U. (1993) Global warming and electricity supply: towards the integration of energy and environment policy in the European Community? DPhil thesis, University of Sussex, forthcoming.

Department of the Environment (1992) *Climate change: our national programme for CO_2 emissions.* Department of the Environment.

Elliott, D. (1992) Renewables and the privatisation of the UK ESI. *Energy Policy* 20(3), pp. 257-268.

Enquete-Kommission (1990) *Schutz der Erde* (Volume 2). Deutscher Bundestag, Referat Öffentlichkeitsarbeit, Bonn.

Fisher, D.(ed.) (1990) *Options for reducing greenhouse gas emissions.* Stockholm Environment Institute, Stockholm.

HM Government (1990) *Our common inheritance.* HMSO, London.

IEA (1987) *Energy conservation in IEA countries.* OECD/IEA, Paris.

Jochem, E.; Gruber, E. (1990) "Obstacles to rational electricity use and measures to alleviate them", *Energy Policy* 18(4), pp.340-350.

VROM (1990) *National environmental policy plan plus.* VROM, Den Haag.

VROM (1992) The Netherlands' memorandum on climate change. *Executive Summary.* Vrom, Den Haag.

10 Success through lack of competition: The political opportunity structure for EC environmental policy in Belgium

Marc Hooghe

Introduction

The environmental policy pursued in Western Europe is dominated by the European Community law and policy. Since 1972, the Community has issued some 200 environment directives, which cover many aspects of environmental protection. The trend toward European integration of environmental policy has already been noted in severallegal works (Kiss 1989; Johnson & Corcelle 1990; Bocken & Ryckbost 1990), but it has received little attention from researchers (Judge 1992). However, the legal approach suffers from serious deficits: lawyers tend to devote their attention solely to one aspect of policy – the issuing of legal texts. The implementation phase of the policy cycle is often neglected in their work. And yet, implementation may be the crucial stage to study the value of EC environmental policy. As Collins and Earnshaw (1992, pp. 213-4) remark:

> "While the processes of initiation, scrutiny, amendment and agreement of EC environmental legislation are important, Community legislation will not be worth the paper it is printed on if policies break down or obligations are not fulfilled at the implementation stage."

If we want to arrive at an assessment of the impact of European environmental policy, the policy cycle as a whole should be studied. The crucial question is whether the member states comply with EC directives, and what impact the directives have on the policy-making process in the member countries (Hildebrand 1992). In this chapter, I will try to answer this question for Belgium: what impact has European environmental policy had on environmental policy?

To arrive at an answer, I will use a slightly amended version of the political opportunity structure approach (Kitschelt 1986). The concept was originally developed to analyse the political impact of non-institutional actors, but, with some slight modifications, it can be used in the case of an institutional actor like the EC. This innovation offers a common methodology to assess the impact of both kinds of actors on the decision-making process. Maybe institutional and non-institutional actors play in a different league, but they certainly play the same game: both try to influence the policy-making process.

The use of this methodology should prevent scholars from making statements on European policy which are too general; its impact should be assessed within an opportunity structure which is specific for each member. The strength of an actor, and its ability to influence decision-making processes, is a relative quality – not an absolute one, it all depends upon the specific power structure on a given policy area.

In the following two sections I will first sketch briefly the actual impact EC environment directives have had in Belgium. From the fourth section onwards, I will seek to analyse these findings, using the political opportunity structure approach.

The Belgian paradox

Belgium suffers from a reputation of being one of the most lax implementors of EC environment directives. In no other country has the Commission of the European Communities identified more infringements of environment directives. Belgium can be considered 'the bad pupil' of European environmental policy. In his influential report on the implementation of European law in Belgium, the Belgian member of the European Parliament Karel De Gucht (1987, p. 125) wrote: "Environment directives are grossly neglected by Belgium, not only by the national authorities, but also by the autonomous regions".

And yet, quantitative research shows that EC directives have a very determining influence on the environmental policy in Belgium, and are responsible for more than half of all environmental measures in the country (Hooghe 1992a). At first sight, this conclusion seems to contradict current opinion. However, this is only an apparent contradiction. Belgium may be a lax implementor of European environmental policy, but this has not stopped the Community from becoming the major force behind the realization of environmental policy in the country. The fact that at least 34 environment directives are not correctly applied should not hide the fact that some 170 environment directives *have* been implemented in Belgium.

A key factor to explain the relative success of the EC, is the lack of competition: the EC can pursue its objectives almost undisturbed, because there are few other powerful actors in the environmental arena. The vested interest groups in Belgium have little interest in the environment, while environmental groups lack the resources which are necessary to influence government policy (Leroy & De Geest 1985).

The apparent paradox becomes clear, if we use the concept of a political opportunity structure. As Kitschelt (1986, p. 58) formulated it: "Political opportunity structures are comprised of specific configurations of resources, institutional arrangements and historical precedents for social mobilization, which facilitate the development of protest movements in some instances and constrain them in others". The main argument is that the possibility of actors to influence a certain political area is to a large extent determined by the overall power configuration of the particular policy area, as it is expressed in terms of resources, institutional arrangements and historical precedents. There is no reason why this should be valid only for non-institutional actors, and not for institutional actors such as the EC. This implies that the influence of European environmental policy is not only determined by its inherent strength or weakness, but also by the relative strength of other actors in the environmental field.

A major problem with EC environmental policy, is that the Community does not have the authority, or personnel, to enforce its directives. To do so, it has to rely on other agencies, which are often not motivated to implement European policy. The situation facing the EC, or any other supra-national organisation, is therefore entirely different from the one facing a government agency, which unites the decision-making phase, and the implementation and monitoring phase in one administrative body. This makes it all the more opportune to use a political opportunity structure approach to study the impact of EC environmental policy: similar to a non-institutional actor, the EC has to rely on government agencies which it cannot control, to secure implementation of the policy it advocates.

The actual impact of European environmental policy

The impact of European environmental policy on environmental policy pursued in Belgium is indeed overwhelming. Earlier research shows that the Community is directly responsible for 58% of all environmental legislation in Belgium and Flanders, the largest autonomous region of the country. No less than 78 of the 134 most important environmental measures, issued between 1980 and 1990 were brought into being by European environment directives (Hooghe 1992a). However, the impact of European directives varies greatly between the various areas of environmental policy, and this becomes clear if we take a look at some of these domains.

The influence is most apparent with regard to legislation on hazardous substances. European involvement in this area has a long histor:, the first directive (67/548/EEC) was issued in 1967. The text of this directive has regularly been brought up to date and has now developed into a very detailed and far-reaching juridical instrument. This comprehensiveness explains why the EC exerts such influence in this area.

The influence of the European Community is least noticeable in the field of nature conservation, where the freedom of scope of the EC is limited by the absence of a clear economic motive. Only in 1979 did the Community issue its first directive on nature conservation, and even now, the number of directives in this area remains limited, and are often incorrectly implemented. To give one example – several member countries, including Belgium, have been convicted by the European Court of Justice for their failure to fully implement directive 79/409/EEC on the preservation of the bird population in the Community. This failure was mainly caused by the lobbying of special interest groups of bird-catchers, hunters, and farmers, which are particularly powerful, in the southern member states.

What becomes apparent in this overview is that it is difficult to make general statements about the impact of European environmental policy. In areas where the EC has issued a comprehensive directive, the European influence proved to be considerably larger than in areas where it has not done so. But the ability of the Community to issue such directives is limited by the willingness of the 12 member countries to cede their authority in this area to Brussels. There is also an historical angle: in areas where the EC already issued directives, the European legislation has had time to influence the national policy. In areas where European involvement is relatively recent, this has not yet occurred.

The quality of European legislation is another important element. Directives with a clear and circumscribed objective are implemented rather well by the member states. But the implementation of some directivès is simply not feasible, because their text is either complicated or ambiguous. This is often a result of the difficult process of forging a compromise within the European Council (Johnson & Corcelle 1990; Collins & Earnshaw 1992). Special interest groups, too, are sometimes responsible for weakening the text of a directive. Via a flourishing lobby sector in Brussels, they exert considerable influence on the decision-making process within the Community (Julien 1990; Andersen & Eliassen 1991). Finally, the success of European involvement is also determined by the number of competitors in an area of environmental policy. If legislators, environmentalists, and the general public voice concrete demands for policy action, the EC will be but one voice in a choir trying to influence policy.

Political opportunity structure

Kitschelt was not the first author to highlight the importance of political opportunity structures in understanding the behaviour of social actors. Eisinger (1973) already used a comparable concept in his study on riots in American cities, and McAdam (1982) explained the rise of black insurgency in the United States in the 1950s and 1960s through changes in the political process, which created opportunities for black protest behaviour. However, Kitschelt, in his study on anti-nuclear movements in four western countries, has used the concept most systemically. Kitschelt's approach has been used to study the impact of West European peace movements (Overby 1990), however his systematization has also received severe criticism (Geier 1990). In its original formulation, Kitschelt's approach was applicable only to social movements, and other non-institutional actors. However, the approach is in principle also valid for institutional actors. In some respects, the situation facing non-institutional actors is completely different from the one facing institutional actors. But these differences should not hide the fact that both of them have the same goal – trying to influence the political decision-making process. In this endeavour, both of them will be confronted with basically the same barriers and difficulties. Therefore, for both of them, a political opportunity structure approach is applicable.

This extension of Kitschelt's approach has considerable advantages: it facilitates the comparison between the impact of both kinds of actors on the decision-making process. Furthermore, the political opportunity structure approach becomes more easily compatible with theories on thè agenda setting process, whereby it is assumed that each item must run an obstacle course before it can gain entrance to the political agenda. Whether the item is promoted by an institutional or a non-institutional acto, does not fundamentally alter that process (Cobb & Elder 1972; Nelson 1984).

Kitschelt identifies three areas where the political opportunity structure may be applied: the emergence and strategy of political actors; the openness of the political system to influence from political actors (input side); and the capacity of the political system to implement the policies which have been agreed upon.

Emergence and strategy

On the level of emergence and strategy, Kitschelt (1986, pp. 61-62) differentiates three variables determining the capacity of social actors to impose themselves as an actor in the political arena. First of all, the actor must be able to mobilize resources; secondly the institutional rules governing the access to the political arena; and thirdly precedents set by earlier, similar, actors.

In the case of the EC, it is clear that the Community controls enormous resources in its struggle to influence the environmental policy in its member countries. First of all, the EC has a juridically established authority: in theory, the member states have no other option but to implement EC directives. If a member state fails to do so, article 169 of the Treaty of Rome empowers the European Commission to start legal proceedings with the European Court of Justice (Hooghe 1993).

As regards the institutional rules governing the access to the decision-making process, there is also an obvious disparity between institutional and non-institutional actors. Whereas a social movement has to enforce itself as a legitimate actor in the political arena, the Community is able to use routine channels to communicate with decision makers. However, one should not view the political system of a country as a monolithic unity. The political system is composed of numerous, often semi-autonomous units. In this case, the EC has routine links with the Ministry of Foreign Affairs in Belgium, which passes the new directives to the ministry responsible for their implementation (Bocken & Ryckbost 1990). However, those ministries are not always motivated to act quickly to implement directives.

The difficulties facing the Community should not be underestimated at the level of historical precedents either. Many areas of competence, which hitherto were considered an area of national sovereignty, are now in the hands of the EC. One might expect the EC to be able to exert its influence in those areas which already have a long history of international involvement, such as economic integration. In contrast, the influence of international environmental policy is fairly recent, and has not yet gained much legitimacy for many decision makers.

Input side structure

Kitschelt differentiates three variables determining the openness of a political system: the number of actors; the balance between legislative and executive branches of government and existing patterns of intermediation.

Kitschelt states that the larger the number of actors trying to influence the decision-making process, the greater the openness of the system will be to outside influence. This hypothesis, however, must be qualified. It can indeed be the case that a large number of political actors will give weight to an issue in political terms, and will induce decision makers to develop a policy on that area. But, on the other hand, a large number of actors can dilute the influence of a single actor, because it has to share the field with other actors, who might make completely opposite demands. Therefore, Kitschelt's hypothesis might be limited to a specific moment in the agenda setting process: when an item gains entrance to the social or political agenda (Cobb & Elder 1972).

In Belgium, the number of actors urging the decision makers to develop a coherent environmental policy is too small to let the environment gain priority on the political agenda (Gobin 1986). One might expect the impetus for developing a more effective environmental policy to come from environmental groups, ecological parties, or the civil service. But in Belgium, none of these actors is powerful enough to push the environment issue to the top of the political agenda. This is mainly due to the closed, neo-corporatist structure of the Belgian political system, whereby access is limited to established, and pillarized special interest groups (Lijphart 1981; Leroy & De Geest 1985). Ecological parties too, have not succeeded in making the environment a top priority. Despite the fact that in the latest general elections they obtained 10% of the vote (Hooghe 1992b), their influence on the political process had thus far remained rather limited. Deschouwer (1989, p.53) writes that "the Belgian Greens have almost no political relevance at the national level".

The closed nature of Belgian policy-making makes it extremely difficult for challenging groups to exert any real political influence. Political structures in Belgium are generally perceived as ossified, and not responsive to social demands (Hooghe 1992c). This means that international organizations are the only remaining possible source for outside pressure on the political system to integrate the environment on the political agenda. The flow of EC environment directives has served as an incitement towards the Belgian authorities to take the environment more seriously.

Kitschelt suggests that the openness of a political system is increased if the legislative branch of government is more powerful than the executive branch. This is indeed the case for social movements, but for the EC, this variable has proved to be irrelevant (Hooghe 1992a). If the Belgian parliament has to take action to implement a directive, this usually takes three or more years, due to the cumbersome procedure in the Lower House and the Senate. If, on the other hand, the government has the authority to implement a directive without the need for approval by parliament, the length of the procedure can vary between four months and six years. Here, a lot seems to depend on the motivation of the individual minister responsible for the implementation. In the frenzy leading to the completion of the internal market on 1 January 1993, both the legislative and the executive branch stepped up their efforts to incorporate European directives more swiftly (Devuyst 1993), but there is no guarantee that this effort will be sustained, now that such a deadline is no longer in sight.

The patterns of mediation between interest groups and the government are the third element shaping the political opportunity structure on the input side. Kitschelt states that where pluralist patterns are dominant, access for new actors will be facilitated. Belgium is traditionally portrayed as having a neo-corporatist political culture (Lijphart 1981), and this makes it more troublesome for new actors to gain access to the decision-making process. This is true both for institutional and non-institutional actors.

Finally, Kitschelt states that openness is constrained if there are no viable procedures to build effective policy coalitions. This barrier is difficult to overcome with the Community – the EC does not seek to build coalitions. The text of its directives is not amendable; the EC is not open to any compromise on its policy goals, and the Community insists on a full implementation of its directives. A non-institutional actor, on the other hand, will find it easier to overcome this barrier, because it has the flexibility to participate in a bargaining process to attract potential coalition partners.

IThe political opportunity structure on the input side seems rather bleak for the EC. From the four variables we have discussed only one proved unambiguously positive – the number of actors in what I called the second arena of the agenda setting process is limited. The three other variables were either not relevant or negative for the EC.

Output side structure

The third level of the political opportunity structure is the output side of the political system. Kitschelt (1986, pp. 63-64) differentiates between three variables on this level: the degree of centralization; government control over market participants; and the relative independence and authority of the judiciary.

Up until recently, Belgium was a highly centralized country. A series of changes in the constitution in 1970, 1980, 1988 and 1993 created autonomous regions. This has led to difficulties for the EC; the European Commission and the European Court of Justice deal only with national governments, work formally with local or regional authorities. Therefore, a high degree of decentralization can facilitate the exertion of influence by non-institutional actors, as Kitschelt states, but makes it more difficult for institutional actors to gain hold on a policy domain.

Kitschelt goes on to suggest that "government control over market participants is a key variable for government effectiveness in many policy areas" (p. 64). But as Kitschelt already indicates: it is difficult to make general statements on this variable. For example, it is relatively easy to monitor the implementation of EC emission standards on car exhausts, because the number of car producers is very limited. The monitoring of the waste directive, on the other hand, is quite troublesome because there are so many actors involved. Therefore, if we want to operationalize this variable, it should be specified for which area it is used.

A third and final variable on the output side, is the relative autonomy of the judiciary. The Belgian constitution grants a large degree of independence to the judicial system but its scope of authority remains limited. Nevertheless, appeal to the judiciary has resulted in some remarkable successes for the environmental movement. In contrast to national courts and tribunals, the European Court has no authority to force the member countries to comply by its verdicts. For example, Belgium has already been repeatedly convicted for its failure to implement earlier verdicts of the Court. If, and when the Maastricht Treaty enters into force, the Court will receive the authority to impose a penal sum on negligent member countries. But as long as this is not the case, the legal proceedings provide only limited opportunities for the EC to enforce its authority.

Conclusion

Students of political decision-making processes have too often taken for granted that institutional actors have no trouble whatsoever in imposing their views. While non-institutional actors have to overcome almost unsurmountable difficulties to gain entrance to the political arena, institutional actors are often portrayed as having automatic and continuous entry. By extending Kitschelt's political opportunity structure approach, to allow it to cover institutional actors too, I have tried to remedy this one-sided representation.

I do not want to imply that the difficulties for institutional and non-institutional actors are identical. It is self-evident that institutional actors have a more privileged position than non-institutional actors. In the case of Belgian environmental policy, it is clear that the impact of institutional actors is ten or more times greater than that of non-institutional actors.

If we apply this framework to the specific case of the European influence on the environmental policy in Belgium, it becomes clear that the differences between the EC and a non-institutional actor, such as an environmental group, are indeed significant. But the two kinds of actors do not belong to two different worlds: the nature of the obstacles they have to surmount is more or less comparable.

Table 2: Political opportunity structure for the European Community and an environmental group as regards environmental policy in Belgium

	European. Commission	Environmental Group
Emergence and strategy		
1. resources	++	--
2. institutional rules	++	--
3. precedents	n.r.	-
Input side		
1a. number of actors, phase I	-	-
1b. number of actors, phase II	+	-
2. legislature	n.r.	-
3. intermediation patterns	-	-
Output side		
1. centralization	-	+
2. market control	n.r.	n.r.
3. judiciary	-	+

++: very positive; +: positive; -: negative; --: very negative;
n.r.: not relevant

It is interesting to note that the opportunity structures of the two kinds of actors are to some extent mirror images. For the institutional actor, the emergence and strategy level is very positive, while this is the level where the non-institutional actor scores worst. For both actors, the input side level is troublesome, while at the output side, the non-institutional actors have a more positive opportunity structure. This indicates that both kinds of actors have an complementary role to play in the realization of a coherent environmental policy. The institutional actor has the financial and legal resources to establish a general framework, but it lacks the suppleness to monitor whether every single administrative or government decision is in accordance with that framework. This is especially the case for an international organization such as the EC, which has to rely on other government agencies to implement its directives. Environmental groups, on the other hand, may not have enough power to influence the legislative process, but their experience and legal action, are important assets when it comes to the implementation of environmental policy. Environmental groups can be better equipped than an institutional actor to fulfil a watchdog function.

It is clear that EC environmental policy has not halted environmental degradation in the Community. At best, it has to some degree slowed down the process. To a large extent, this is due to the inherent weaknesses of the European policy-making process (Liberatore 1991; Hooghe 1993; De Grauwe 1993). European legislation on the environment is often bureaucratic, relies too heavily on top to bottom approaches, sometimes lacks coherence and transparence, and in some cases application is simply not possible. The EC often resembles a slow moving mammoth, ill-equiped to react swiftly and accurately to environmental threats. This can partly be remedied by beefing up the efforts to establish a coherent EC environmental policy. Nevertheless on certain accounts, the political opportunity structure for a large, institutional and supra-national actor like the EC will remain negative. One should not expect every solution to come from a group of Brussels bureaucrats. Preserving a sound environment requires the participation of a range of social actors, and this is, indeed, the basic philosophy behind the Fifth Action Plan for EC environmental policy (CEC 1992a). The role of non-governmental environmental organisations is crucial in this aspect, as has been recognized by the recent United Nations Conference on Environment and Development (UNCED). Non governmental organisations have a role to play, not only in the decision-making phase of the policy cyle, but also in the phase of implementation and monitoring. Combining the frequently complementary strengths of institutional and non-institutional actors in terms of their different political opportunity structure, can only lead to a more effective and vigorous environmental policy.

References

S Andersen & K Eliassen (1991), 'European Community Lobbying', *European Journal of Political Research*, 20, 2, 173- 187.

H Bocken & D Ryckbost (eds., 1990), De voorberei-ding en tenuitvoerlegging van Europese richtlijnen inzake de bescherming van het leefmilieu. Brussels: *Story Scientia.*

Roger COBB & Charles ELDER (1972), Participation in American Politics. *The Dynamics of Agenda-Building.* Baltimore: John Hopkins University Press.

CEC (1992a), *Towards Sustainability*. Document COM(92) 23final, 3 vol. Brussels: Commission of the European Communities.

CEC (1992b), *Ninth Annual Report on Commission Monitoring of the Application of Community Law.* Doc. COM(92) 136final. Brussels: Commission of the European Communities.

K Collins & D Earnshaw (1992), 'The Implementation and Enforcement of European Community Environment Legislation', *Environmental Politics,* 1, 4, 213-249.

K Deschouwer (1989), 'Belgium: the "Ecologists" and "AGALEV"', pp. 39-53 in F Müller-Rommel (ed.), *New Politics in Western Europe.* Boulder: Westview Press.

Y Devuyst (1993), 'De omzetting van EG-richtlijnen in de Belgische rechtsorde en de europeanisering van de Belgische politiek', *Res Publica*, 35, 1, 39-54.
P Eisinger (1973), 'The Conditions of Protest in American Cities', *American Politicial Science Review*, 68, 11-28.
K Geier (1990), 'Peace Movements and the Struggle Within', *West European Politics*, 13, 275-279.
C Gobin (1986), 'L'état belge et la problématique de l'environnement', *Courrier Hebdomadaire du CRISP*, nr. 1109.
P De Grauwe (ed., 1993), *Economic Instruments in EC Environmental Policy*. Brussels: Centre for European Policy Studies.
K De Gucht (1987), *De toepassing van het Europees gemeenschapsrecht door België*. Document 86-87/20-739/1. Brussels : Chamber of People's Representatives.
P Hildebrand (1992), 'The European Community's Environmental Policy', *Environmental Politics*, 1, 4, 13-44.
B Hogwood & L Gunn (1984), *Policy Analysis for the Real World*. Oxford: Oxford University Press.
M Hooghe (1992a), 'De invloed van het internationaal milieubeleid op de totstandkoming van het milieubeleid in Vlaanderen', *Res Publica*, 34, 1, 3-23.
M Hooghe (1992b), 'The Green Parties in the Belgian General Elections of 24 November 1991', *Environmental Politics*, 1, 2, 287-292.
M Hooghe (1992c), 'Wat is er verkeerd met België ?', *De Nieuwe Maand*, 35, 9, 9-13.
M Hooghe (1993), 'Too Little, Too Slow. The Impact of European Environmental Legislation on the E.C. Member States', *Planologisch Nieuws* (in press).
S Johnson & G Corcelle (1990), *The Environmental Policy of the European Communities*. London: Graham & Trotman.
D Judge (1992), 'A Green Dimension for the European Community ?', *Environmental Politics*, 1, 4, 1-9 (Special Issue on 'A Green Dimension for the European Community').
B Julien (1990), 'Euro-Lobbying invades Berlaymont', *European Affairs*, nr. 3, 28-33.
A Kiss (1989), *Droit international de l'environnement*. Paris: Pedone.
H Kitschelt (1986), 'Political Opportunity Structures and Political Protest', *British Journal of Political Science*, 16, 1, 57-86.
P Leroy and A De Geest(1985), *Milieubeweging en milieubeleid*. Antwerp: Nederlandsche Boekhandel.
A Liberatore (1991), 'Problems of Transnational Policy Making. Environmental Policy in the European Community', *European Journal of Political Research*, 19, 2/3, 281-305.
A Lijphart (ed., 1981), *Conflict and Coexistence in Belgium*. Berkeley: Institute of International Studies.

D McAdam (1982), *Political Process and the Development of Black Insurgency.* Chicago: University of Chicago Press.
B Nelson (1984), *Making an Issue of Child Abuse. Political Agenda Setting f or Social Problems.* Chicago: University of Chicago Press.
L. M Overby (1990), 'West European Peace Movements: An Application of Kitschelt's Political Opportunity Structures Thesis', *West European Politics*, 13, 1-11.

Part IV

URBAN ENVIRONMENTAL PROBLEMS: EAST AND WEST

11 Urban environmental pollution and child health in Houston, USA: The links to economic growth

Judith Chernajovsky

Introduction

Economic development and environmental change are intricately linked. Unfortunately, they are often treated as though separate, and so is the analysis of environment and development . Solutions to the problems remain piecemeal[1] .

This chapter argues that massive capital accummulation and industrialization may bring affluence, but also can bring environmental deterioration and disease. This position is examined in Houston, Texas, U.S.A. with a focus on the impact of air pollution on child health. The core of the paper consists of a comparative study of two residential areas in the city , each affected by different levels of air pollution. Similar environmental child health risks exist in other large cities in the developed world, in the U.S.A. and Europe, highlighting the significant negative implications that economic development can have on health.

Sources of toxic emissions have multiplied since World War II. This has resulted in a previously unknown magnitude of air pollution. Back in the 1950s, Kapp[2] argued that the social cost of atmospheric pollution goes far beyond losses resulting from the more rapid deterioration of building materials, metals and paint coatings. He thought that the loss of day-light and of ultraviolet light might have had the most injurious effects upon human health and that the damage to persons was also caused by the inhalation of smoke-polluted air. Kapp's suspicions and views were based on relatively rudimentary scientific and technological assumptions when compared with the state of the art forty years later. In spite of this, studies -including this one- have shown that Kapp's ideas from the 1950s are still relevant and, perhaps, more pertinent than before.

Urban environmental consequences of economic growth

The study of any environment-society relationship in Houston cannot be isolated from political decisions about economic growth. In Houston, expansion of the oil-related industry has been at the heart of the city's economy during the last 50 years.

Although the economy in Houston is multi-sectoral, oil related industries are still at the core of economic activity. Many aspects of growth in the city, such as population, building, industry, and not least, environmental pollution, link to oil-related activities. Since 1850, the rate of population growth per decade in Houston has averaged 66 per cent, a level unparalelled over such a long period among American cities. In 1890, it was ranked as 112th-largest U.S. city. In 1980, Houston was the fifth-largest U.S. city. By 1983 it was the fourth-largest. Natural increase, migration and annexation all contributed to this growth.

Over the last few decades, Houston has dramatically expanded both horizontally across the land and vertically into the air. Industrial and finance capitalists have been the driving forces behind growth in Houston. Houston epitomises cities which have undergone rapid development since World War II. The pre-eminent role of oil as a source of energy in the American economy, pre- and post- World War II, greatly contributed to the rapid growth of Houston as petroleum processing centre. Oil was already being produced in Houston and there was a solid city infrastructure due to the commercial character of the city. Moreover, the state subsidized chemical facilities during and after the war. Such was the growth and the success of the petrochemical industry in Houston that, by the 1960s, the city became the 'energy capital' of the U.S.A. Within Houston, large petrochemical plants located near the Ship Channel, where they could easily obtain feedstocks by pipeline and had ready access to waterborne transport[3].

Today, the Houston Ship Channel and the Port of Houston are the main water routes through which oil reaches, is processed and leaves Houston. Most manufacturing facilities, refineries and the port are in the east side of the city, along the ship channel. This is the main commercial artery in the economic life of Houston. There are hundreds of industrial plants spewing pollutants into the air. Many residential areas are located along some sections of the Ship Channel. Some neighbourhoods are periodically affected by uncontrollable release of pollutants, or by accidental hazardous spills.

It should be added that the road system in Houston is one of the most extensive in the U.S.A. Numerous highways, together with the absence of any commitment to a significant mass transit system, leave Houston a completely car-dominated city. In addition, more than two million cars contribute to the smog. Car transport is an additional major source of air pollution in Houston. Nevertheless, in contrast to other newly developed cities, air pollution in Houston is related more to industrial plants than to cars[4].

Another environmental consequence of rapid economic growth in Houston is the problem of toxic waste dumping. Cheap and generally unrestricted toxic waste dumping is linked to industrial profitability. Expanding industrial investment since the 1920s has caused rapid growth of toxic waste, particularly during the postwar period. In the Houston-Galveston-Gulf Coast area industries produced three million tons of toxic waste each year in the early 1980s. In addition, billions of gallons of contaminated water were being dumped into local wells. Moreover, much hazardous waste flows to, or through, Houston by lorry, airplane, train and pipeline. Also, periodically there are a major transport and plant disasters causing dangerous spills and gas leaks[5].

In summary, the petrochemical and other industries, a growing fleet of cars, a lack of public transport, a flourishing toxic waste business and a weak regulatory environment, all contribute to the high levels of air pollution found in the 'wealthy' city of Houston. It may be said that society functions in the city of Houston to amass monetary wealth, but in so doing, undesirable pollutants are inserted into the environment. Thus, air pollution is the outcome of policies which have continuously encouraged capital accummulation and urban growth while disregarding the major effects on other vital aspects of city life.

The comparative study and the situation on the ground

During the last decades, special efforts have been made by epidemiologists, medical geographers, environmentalists and sociologists to ascertain the extent and type of health risks involved in exposure to air pollutants. Unfortunately, many of these studies have looked at health risk and pollution as an isolated issue. The present study aims to make a contribution to this literature but also aims to extend it by looking at the relationships between this issue and the political economy. This paper examines empirical information from the Air Pollution and Child Health Survey[6] carried out in Houston, and assesses results in relation to political economy of growth in Houston.

The empirical information in this paper is based on a comparative study of two different areas of Houston, with different levels of air pollution. These are (i) the more polluted area, in the *Ship Channel* , a residential area located in the proximity of large industrial plants and the Port of Houston; and (ii) the less polluted area, in the *South-West*, a residential area located about 20 miles away from industrial and port facilities. Two different sets of information are employed. One set originates in primary data from the household Air Pollution and Child Health Survey, conducted by the author in Houston. The household survey was carried out in 1990 (n=300 households) in both areas to establish frequency of pollution odours, child health status and household socio-economic attributes. The survey provides a unique source of information from various sources including: the extent and type of child illness in Houston, the frequency and character of air pollution in the study areas as perceived by respondents, household socio-economic status (income, dampness) and demographic composition (numbers of children per household).

In addition, levels of toxic emissions (sulphur dioxide and ozone) are examined. This second set of information comes from secondary sources, the City of Houston Bureau of Air Quality Control -BAQC-[7] and it is useful for assessing the levels of sulphur dioxide and ozone in the polluted and less polluted areas.

Statistical analysis of primary data was carried out by using SPSS. The extent of child ill-health and of socio-economic conditions in low and high-income homes in both, the polluted and less-polluted areas was first analyzed by means of cross tabulations and analysis of variance. Chi-square tests were carried out on the significance of the association between incidence of child diseases by household status and area of residence.

Air pollution in the urban environment

Most of the major industrial manufacturing plants in Houston are placed on the east side of the city, extending about 25 miles along and south of the ship channel: chemical plants, oil refineries, other oil related facilities, steel and other metals facilities. It is in these surroundings that the polluted residential study area is located. Two types of air pollution measures are examined, sulphur dioxide and ozone levels as measured by the BAQC and perception of pollution as defined by local residents.

Only ozone levels exceed the NAQQS (*National Ambient Air Quality Standards* set by the U.S.A. Environmental Protection Agency) in Houston. Air pollution in Houston has been measured as the second worst in the U.S.A. after the city of Los Angeles[7]. It has been classified a non-attainment area for O_3 concentrations[8].

Within Houston, however, although the other toxic concentrations are within the health safety limits, there is a pronounced spatial variation in the distribution of emissions in the atmosphere across the city. Although the annual average concentrations of sulphur dioxide in the polluted area (0.13 PPM) does not exceed NAAQS (0.14 PPM), this level is high enough to suggest that the quality of the air in Houston is poor in some parts of Houston. Monthly concentrations of sulphur dioxide in the Ship Channel range from 0.09 in April to 0.22 PPM in January 1989, the last being well above the health safety NAAQS. The annual average sulphur dioxide measures in the less-polluted area, on the other hand, only reaches just about one-half (0.07 PPM) that in the polluted area. The lowest level registered in the less polluted area is 0.04 and the highest 0.17 PPM.

Levels of ozone in Houston, which are very high[8], also oscillate according to monitoring station and the month of the year. Only one month, February, shows low concentration but relatively high levels of ozone are registered for the rest of the year. The annual ozone average in the polluted area reaches 0.16 PPM but the monthly average levels broadly oscilate between of 0.06 PPM (lowest) and 0.25 PPM (highest). This measure exceeds the 0.12 PPM NAAQS for ozone established by EPA. The annual concentration of ozone in the less-polluted area is lower than in the polluted area. Having said that, ozone pollution in the South-West is very high as well, 0.15 PPM. In 1989,

Houston exceeded the EPA national standard of 0.12 PPM for ozone gases in the atmosphere for more than 1 hour, once a year, for 45 days a year.

Figure 1.Sulphur dioxide and ozone concentrations in the polluted (Ship Channe)l and in the less-polluted (South-West) residential areas, Houston, 1989

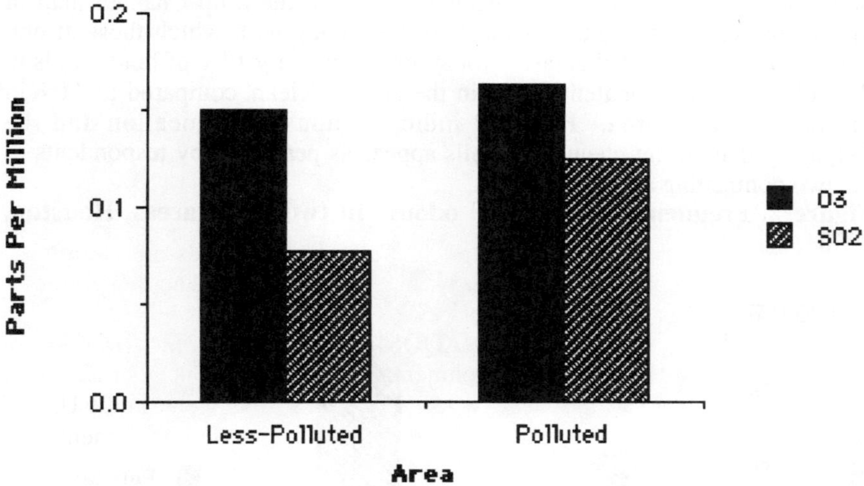

Source: data adapted from The City of Houston Bureau of Air Quality Control , 1989.

Levels of pollutant concentration measured of different monitoring stations draw attention to spatial variation in the distribution of toxic emissions around the city. The variation found in the environmental conditions within the city leads to two further questions related to the empirical purpose of this study: how do local residents perceive air pollution and what is the extent to which the different concentrations of air pollutants in the air might be hazardous to child health.

Residents' perception of air pollution

A major distinguishing feature of atmospheric pollution in Houston is odour. This can be best described as a distinctive, sulphuric smell which is present most of the time and can be pungent and stingy according to the time of the year and the area of the city. Odours have actually been used to detect accidents, the bases of residents' complaints to the authorities, and sometimes, to investigate sources of emission[5]. The smell originates in emissions released into the air by the industrial plants situated along the Ship Channel. Apart from petrochemical and other industrial sites, a number of toxic waste disposal plants contribute to this environmental pollution. The local wind pattern has an additional effect on the distribution of suspended particulates in that the prevailing south-easterly winds extend the patterns of pollution toward the north-west and, in the three months of winter, toward the south-west[10]. High temperatures and proximity to the Texas City-Galveston area are also factors which contribute to air pollution potential[8].

In the polluted Ship Channel residential area, a majority, -83%- of the surveyed households, complained that there are usually strong chemical odours in the air. This compares to 29% of households in the less polluted South-West. Moreover, 36% more households in the Ship Channel than in the South-West complained of the high frequency with which these strong odours invade residential areas. Not surprisingly, only 13% of households in the polluted area evaluated the air in the area as 'clean' compared to 41% in the less-polluted area. Figure 2 indicates household location and the frequency that strong chemical smells appear as perceived by respondents in the two contrasting areas.

Figure 2. Frequency of chemical odours in two study areas, Houston, 1990

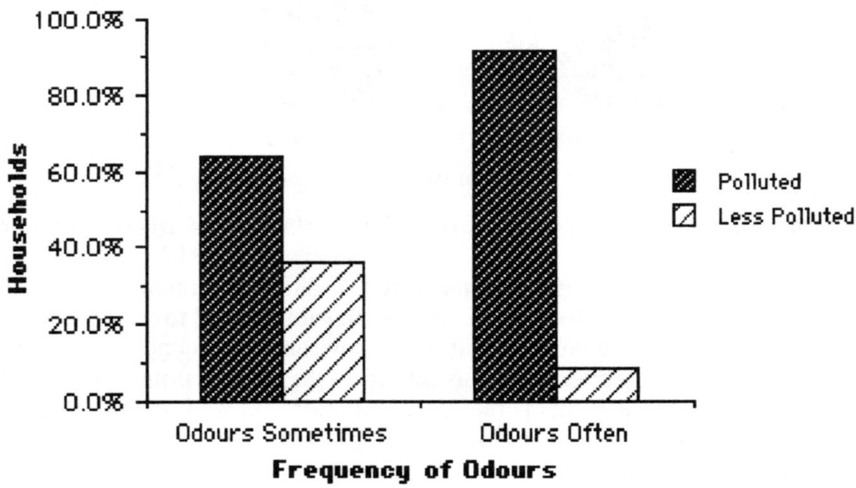

Source: Air Pollution and Child Health Survey, Houston, 1990.

Perception of air pollution by local residents in the two different residential areas gives a clear evidence of the spatial variation and uneven distribution of toxic emissions within the city. These findings clearly agree with the higher concentrations of sulphur dioxide and ozone measured in the polluted area.

Air pollution is known to affect health

The *short-term* effects of extreme conditions of air pollution are well documented. In the earliest part of this century, episodes of acute, severe, particulate air pollution in European and American places (Belgium, 1930;

particulate air pollution in European and American places (Belgium, 1930; Pennsylvania, 1948; New York, 1953; and London, 1948, 1952 and 1956) led to rises in mortality chiefly because of increased deaths from pneumonia and cardiovascular disease [27]. However, the *long-term* health effects of severe and of non-severe episodes of different concentrations of pollutants in the air is not completely known.

Morbidity, rather than mortality, seems likely to be a more sensitive outcome measure in studies of relatively long-term and high, but not extreme, levels of air pollution. Indeed, episodes of extremely hazardous pollution days have decreased considerably in American and European cities. The atmospheric conditions most often reigning in Houston exhibit constant low to medium levels of air pollution with rather occasional extreme carbon monoxide and ozone trends. During a ten years period -1981-1990, according to the Environmental Protection Agency, there have been several 'unhealthy' days every year, few 'very unhealthy' and 'no hazardous' days reported [8]. Most of the days within this period has been classified between 'moderate' and 'good', and are, therefore, within the health safety limits.

Various studies, carried out in highly industrialized big cities in Europe, have demonstrated that air polluted environments contribute to the development of several pathological processes in humans. A 1979 study, pooling results from eight European countries (Czescholosvakia, Denmark, Greece, the Netherlands, Poland, Romania, Spain and Yugoslavia) on chronic respiratory diseases in children in relation to air pollution, reported a close association with air pollution and various respiratory indices in children [12]. This is in agreement with more recent investigations in Barcelona [13], Switzerland [14], Austria [15], France [16], Finland [17] and Italy [18] on the importance of low level pollution on child respiratory conditions. Overall, although the level of contamination in these studies usually does not exceed the established 'safe limits', increases in air pollution have been found to be correlated to the frequency of symptoms.

Respiratory morbidity has been found to be higher in all ages, and ventilatory function in children was poorer in more polluted cities compared with relatively nonpolluted rural areas [19]. Indeed, high levels of sulphur dioxide and particulate matter have been used as predictors of hospital admissions for acute respiratory disease in children and adults [11]. Changes in the biochemistry of the lungs have been documented at ozone levels well below international safety limits [20] and are associated with an increased risk of cough and respiratory illness.

Variation in child illness in Houston

To begin with, the proportion of households with children suffering from any health symptom is very high, and perhaps unexpected in one of the richest cities in the U.S.A. Indeed, 68% of all the households of the sample reported children with ill health. An overwhelming proportion (96%) of health disorders which afflict children in the study areas are of respiratory origin breath).

In general the polluted residential area, 73% of the households reported ill health in their children; this compares to 63% in the less-polluted area (x2=2.87; P=.09). Figure 3 shows the proportion of households with healthy and with ill children in each area .

Figure 3 Child health variation, Houston, 1990
Source: Air Pollution and Child Health Survey, Houston, 1990.

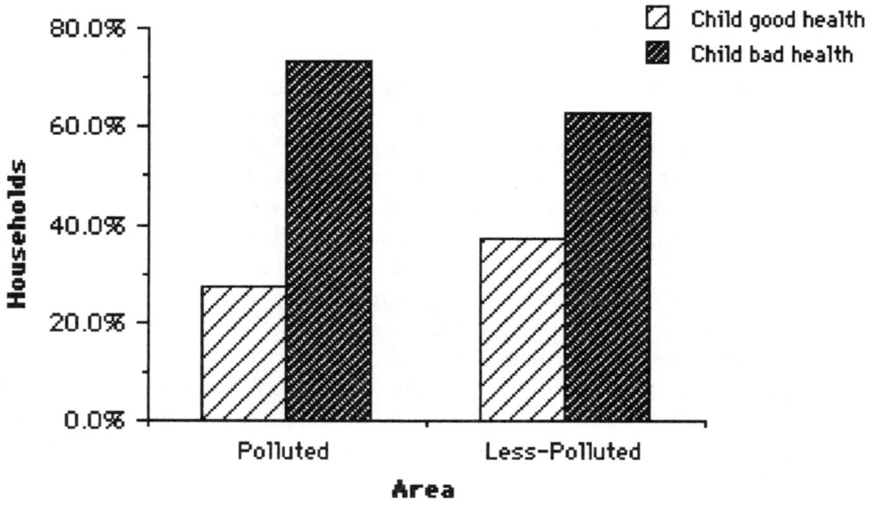

The relative risk of suffering from these conditions was further examined by controlling for income, number of children and dampness in the household. Altogether, more low than high income households (14% more, P=.011, x2=6.37) have ill children. Overall, poor households in the polluted Ship Channel and the less polluted South-West show similar social class and house conditions (although poor households are slightly less well off in the less polluted than in the polluted area). The average number of children per household is similar in both areas and also it is the percentage of houses with dampness (16% in the Ship Channel and 15 % in the South-West). The percentage of households with child illness in low income households is similar in the polluted and in the less polluted residential areas (72% in the Ship Channel and 77% in the South-West).

However, and significantly, child illness in high income households is substantially greater in the polluted than in the less polluted area (73% compared to 49% respectively). Unexpected large numbers of households with ill children cannot be correlated with indicators of class or income. Rather location of the residence is a main factor which determines this particular distribution of child ill health in the study population. The particular distribution of child ill health in the study population. The variation of child health according to household geographical location is indicative evidence of the effects of airborne pollution.

Excess child illness in the polluted areas in the Ship Channel correlates thus with two important examined factors, both of which are present in the same area. These are higher concentrations of sulphur dioxide and ozone and more frequent unpleasant and pungent odours. There is thus a strong relationship between the prevalence of respiratory conditions in children, spatial location of the household and the quality of the air.

Conclusion

Excess child illness has been found in the residential area with more frequent and strong air pollution odours and with more hazardous concentrations of O_3 and SO_2. Neither social class nor demographic information alone are sufficient to explain this variation. The evidence is that the incidence of child illness is associated with the household location in relation to sources of toxic emissions.

Air pollution in the urban environment in Houston is one of the important deleterious effects of pro-growth policies and is associated with massive capital accummulation. Hazardous levels of O_3 and SO_2 in the urban environment is attributable to industrialization over large areas of land and water. Often, the physical environment's own regularities and laws are ignored in profit-centred investments and decisions. Within a given natural environment, the rate of economic growth is linked to the velocity of capital growth and investment. When the capital influx is unrestrained, as in the case of the city of Houston, the absorptive capacity of the social and environmental systems can be quickly overwhelmed. This results in a higher than expected incidence of ill health - in this case, for children suffering from respiratory diseases.

Economic growth has been widely recognized to have contributed to the improvement of the health status of the population. Infant mortality rates in Europe and in the U.S.A. have decreased over the past four decades from nearly 30 to 13 death per 1,000 live births in Europe [21] and from 30 to approximately 9.9 deaths in the U.S.A.[22]. However, many of the effects of economic growth on the urban environment, and ultimately on human beings, are -in reality- more harmful than beneficial. Air pollution, associated with industrialization and morbidity risks, is one such case and it has been a major social cost. In Houston, economic growth is linked to international and national capital accummulation and one main manifestation of successful growth has been the expansion of the oil related industry [23]. As mentioned at the beginning of the paper, petrochemical industries are main sources of harmful pollutants in the sky of Houston .

In summary, examination of the political economy is essential when looking at different aspects of the urban environment and of economic development. Research should consider the political and economic setting of environmental and health issues, not only by juxtaposing them but by relating them in a causal manner. Moreover, public health indicators provide a gateway to other urban environmental problems related to the way our society functions. Results from this type of research should be useful and informative for political decision-making processes. Further research is necessary on the interrelationship between the economic growth of cities, the environment and social health.

References

1. Bartelmus, Peter (1986). *Environment and development.* Allen and Unwin.
2. Kapp, K.W. (1950). *The Social Costs of Private Enterprise.* pp-71-79.
3. Greater Houston Partnership (1991). *Houston Facts* 1990.
4. Feaguin, Joe R. (1988). Free Enterprise City. *Houston in Political and Economic Perspective*. Rutgers University Press.
5. In *Houston Chronicle*: Byars, C. 'Portion of Highway closed by gas leak' 30 September 1990, p. ; Robinson, J., 'Short-staffing cited in failure to detect spill', 30 September 1990; Perry, E.L. and Fair, K., 'Toxic blaze forces up to 1,000 to flee', 30 September, 1990.
6. *Air Pollution and Child Health Survey*, Houston, 1990.
7. Bureau of Air Quality Control (1991). *1989 Annual Data Report.* City of Houston.
8. EPA *National Air Quality and Emissions Trends Report, 1990.* November 1991
9. Texas Air Control Board (1992). *Texas Air Quality Compared to National Levels 1990.*
10. Bureau of Air Quality Control (1991). *1989 Annual Data Report.* City of Houston, Department of Health and Human Services.
11. Graham, Neil M. (1990). The epidemiology of acute respiratory infections in children and adults: a global perspective, *Epidemiologic Reviews*, 12: 149-178.
12. WHO Reports (1979). *Chronic Respiratory diseases in Children in Relation to Air Pollution.* Regional Office for Europe, World Health Organization, Copenhagen.
13. Mallol, Jordi, Nogues, Maria Rosa (1991). Air pollution and urinary thiother excretion in children of Barcelona. *Journal of Toxicology and Environmental Health*, 33:189-195.
14. Rutishauser, M., Ackermann, U., Braum, Ch., Gnehm, H.P. and Wanner, H.U. (1990). Significant association between outdoor NO_2 and respiratory symptoms in preschool children. *Lung* . Suppl: 347-352.
15. Zwick, H., Popp, W., Wagner, C., Reiser, K., Schmoger, J., Bock, A., Hekner, K., Radunsky, K. (1991). Effects of ozone on the respiratory

health, allergic sensitization, and cellular immune system in children, *American Review of Respiratory Diseases*, Nov, 144 (5): 1075-1079.

16. Boussin, G., Leloup, M., Cayla, F., Giroux, M., Pous, J. (1990). Symptomes respiratoires chez les enfants dage scolaire a Toulouse. Influence de la pollution atmospherique, *Archive France de Pediatrique*, May, 47(5):351-6.
17. Jaakkola, J.J., Vilkka, V., Marttila, O., Jappinen, P., Haahtela, T. (1990). The South Karekia Air Pollution Study. The effects of malodorous sulfur compounds from pulp mills on respiratory and other symptoms, *American Review of Respiratory Disease,* Dec. 142:1344-50.
18. Forastiere, F., Corbo, G.M., Michelozzi, P., Pistelli, R., Agabiti, N., Brancato, G., Ciappi, G., Perucci, C.A. (1992). Effects of environment and passive smoking on the respiratory health of children, *International Journal of Epidemiology* Feb, 21(1):66-73.
19. Toyoma, T. (1964). Air pollutions and its effects in Japan. *Archive of Environmental Health,* 8, 153-173.
20. Read, Robert, Read, Cathy, 1991. Breathing can be hazardous to your health. *New Scientist*, 23 February:34-37.
21. World Resources 1988-1989 (1988). A Report by the World Resources Institute, International Institute for Environment and Development, United Nations Environment Programme. New York.
22. Dockery, d.W., Ware, J.H., Ferris, B.G., Speizer, F.E., Cook, N.R. and Herman, S.M. (1970). Change in pulmonary function in children associated with air pollution episodes. *Journal of Air Pollution Control Association.*, 32: 937-942.
23. Shelton, Beth A, Rodriguez, Nestor P., Feagin, Joe R., Bullard, Robert D., Thomas, Robert D. (1989). *Houston. Growth and Decline in a Sunbelt Boomtown.* Temple University Press. Philadelphia.

12 Air pollution management in the former USSR

Maria Shahgedanova and Timothy Burt

Introduction

"The strictness of Russian laws is relaxed by non-obligation to carry them out".

Saltykov-Shedrin, Russian writer, 19 century.

Industrialisation came rather late to the Soviet Union, and as a result high priority has been given to the rapid industrial development and modernisation of the economy. Lenin's statement of 1920, "Communism is Soviet power plus the electrification of the whole country", illustrates this commitment to modernisation. The introduction of the first Five Year Plan in 1928 launched a massive drive to industrialise the Soviet Union regardless of the cost in terms of wasted resources, effort, and lives. Since then, the campaign to transform the USSR into a modern industrial society has always overridden environmental considerations.

The official Soviet approach of always demonstrating a commitment to the protection of nature, has resulted in the development of environmental legislation and the creation of the numerous environmental agencies and departments. It is not the lack of legislation but the difficulty or impossibility of implementing it that created an environmental situation which can only be characterised as disastrous (table 1). This difficulty resulted from the vagueness of the laws, the diverse functions of the agencies in charge of monitoring and a lack of co-ordination between these various agencies. Underlying all this has been the submission of environmental concerns to production quotas.

During the last few years, environmental issues have become of increasing political importance in the former USSR as public awareness of the scope and effects of problems has been growing. Air pollution, being one of the major aspects of the environmental degradation, is now receiving more and more

Table 1. Soviet Cities with the Highest Ambient Concentrations of Dust, SO2, CO, NO2 in 1990 (μgm-3)

City	Pollutant	Annual mean	Daily max	Source of Pollution
Kutaisi (Georgia)	Dust	1250	14800	Track plant,constr'n materials prod'n
Bezmein (Turkmenia)	Dust	1030	5600	Cement prod'n
Norilsk (Eastern Siberia)	SO_2	140	8400	Copper-nickel smelter
Stepnoe (Volga region)	SO2	126	8890	Gas processing
Erevan (Armenia)	NO2	146	980	Non-ferrous metallurgy, chemical industry, power plants, transport
Erevan (Armenia)	CO	6000	46000	Transport
Gorlovka (Ukraine)	NO2	125	660	Ferrous metallurgy

attention; there is a growing number of publications documenting health problems in various cities which are believed to be connected with air quality. Probably the best indicator of the critical environmental situation is the fact that 20% of the population lives in 103 cities where the level of at least on major air pollutant has exceeded the maximum permissible level by 10 times or more on at least one occasion between 1987-1990 (Bond et al., 1990). The total emission of air pollutants in the USSR in 1988 was about 79% of the US total amount. Considering that the Soviet GNP is only some 54% that of the US, this means that the Soviet Union generates 2.5 times more pollution than the USA per unit of GNP (Bond et al., 1990).

In this chapter an attempt is made to analyse the reasons for the Soviet air pollution crisis, to look at the most recent changes in approach to the problem and to forecast its development in the near future.

Environmental legislation and air pollution management

The Soviet record of environmental legislation does not compare unfavourably with the experience of Western industrialised nations; environmental management in the Soviet Union has a broad legal framework. Its expression can be found in the fourth Soviet Constitution adopted in 1977, where it is specified in Article 18 that "necessary steps will be taken to protect and make scientific, rational use of the land and its mineral and water resources, and the plant and animal communities, to preserve the purity of air and water, to ensure reproduction of natural wealth, and to improve the human environment". Further Article 42 stipulates that the citizenry has a right to healthy human environment, and Article 67 obligates them to assume a responsible attitude toward the natural environment and states that "Citizens of the USSR are obliged to protect nature and conserve its riches". These constitution provisions are highly commendable, but do not automatically produce environmental improvements. The quality of the natural environment is more highly correlated with the degree of success achieved in carrying out actual environmental laws. The first example of air quality legislation was a 1949 resolution "On Measures in the Struggle Against Pollution of the

Table 2. Air Pollution Legislation, Laws And Acts in the USSR

Date	Title	Main points
1949	Resolution "On Measures in the Struggle Against Pollution in theAtmosphere and Improving the Sanitary-Hygienic Conditions of Populated Areas"	The State Sanitary Inspectorate was formed to monitor harmful emissions. Provisions for the purification of waste gases were declared to be compulsory.
1951	First list of ambient air standards approved	
1956/63	General environmental laws passed by all the Union republics	
1963	First air pollution monitoring sites established in the USSR	
1969	Resolution "On Improving of Planning and Construction Works"	Planning and constructing of industrial and other sites with consideration of environmental impact including air quality
1970	Standards, norms and methods of determination of lead content in petroleum established	
1972	Resolution "On Strengthening Environmental Protection and Improving of Managing Natural Resources"	
1972	Committe on Hydrometeorology was formed	Monitoring of water and air pollution were delegated to this agency
1972	Air pollution monitoring network established throughout the USSR	
1974	State Inspection for Control of Gas Cleaning and Dust-Catching Equipment established	
1976	Instructions on settlement of conflict in environmental protection published	
1977	Resolution "On Court Environmental Protection Practice"	
1978	Resolution "On Additional Measurs to Strengthen Environmental Protection And Natural Resource Management"	Standards on harmful industrial emissions were adopted
1978	Committee on Hydrometeorology became State Committee on Hydrometeorology and Environment	More functions on environmental control were delegated to the Committee
1980	Fundamental Law on Air Pollution Control	
1981-82	Laws on Air Pollution Controll passed in all the Union republics	
1984	State Committee on Hydrometeorology and Environment was obliged to conduct environmental expertise and issue emission permits	
1988	State Committee for Environmental Protection and Rational Use of Natural Resources was formed	Centralisation of environmental control and management

Atmosphere and Improving the Sanitary-Hygienic Conditions of Populated Areas" (Zeigler, 1987; table 2). This resolution prohibited for the first time in the USSR the construction or renovation of an entire range of industries that did not make provisions for the purification of waste gases; it instructed the Ministry of Higher Education to develop courses on gas and dust trapping; other ministries were to work on purification methods. It also set up a State Sanitary Inspectorate attached to the USSR Ministry of Health to monitor harmful emissions.

Two years later, in 1951, the Air Quality Management Strategy for air pollution control was adopted. The Soviet Union was the first country to propose standardised evaluative criteria for environmental protection in the form of "maximum permissible concentrations" (Russian abbreviation PDK) of harmful substances. PDKs designate optimum levels of ambient air quality which do not affect people's health and do not cause any pathologcal disorders or diseases over the long-term. In 1951, maximum permissible concentrations for 10 air pollutants were established (Zeigler, 1987; Jancar, 1988); in 1981, there were nationally approved PDKs for some 200 individual air pollutants and their combinations. The PDK norms are the same throughout the USSR; stricter regulations may be established in certain regions. For example, within recreation areas a norm of 0.8 PDK is approved. In this respect, the US practice of designation of air quality control regions with different limits throughout the country on the basis of meteorological and topographical factors, urban-industrial concentrations, jurisdictional boundaries, and other factors relevant to effective implementation of air quality standards (Tomany, 1975) seems to be more reasonable (although more complicated), as both in the USA and in the USSR the diversity of natural and socio-economic conditions provide the necessity for different approaches to the environment.

Table 3 compares demonstrates ambient air quality standards for the USSR, the USA and the EEC. In most cases the Soviet standards are of such severity and diversity, that compliance with them is hardly attainable in practice. When a standard is currently unattainable, it is supposed to represent a direction for future enforcement or a guideline for future research in control technology. In this case it works only as a demonstration of the commitment to environmental protection rather than a set of obligatory principles. Partly due to this point, the legislation fails to regulate pollution problems: the laws are so strict that it is often almost impossible to carry them out. That means that many of the polluters will use this reason to cover their insufficient efforts to meet the standards. This approach would possibly work in a highly developed type of community, which recognises the importance of the problem; however, the Soviet society, which according to the communist doctrine is supposed to be one, does not correspond to this image in practice.

The most serious attempt to resolve pollution problems with the help of legislation was made in 1981, when the Air Quality Law (the first national law on air pollution) was passed. This law - an analogue of the American Air Quality Act (1967) and Clean Air Act (1970) or the British Clean Air Act (1956) - took a broad approach and quoted plans for "protecting the air basin as part of the state plans for the economic and social development of the USSR, by the exercise of the state control of air protection, and also by the establishment of ceilings on permitted discharges from the possible sources of pollutants". The basic concept was to establish standards for emissions from all stationary sources and for means of transportation. Emissions by stationary sources were to be individually authorised by state agencies. In practice, these standards known as "maximum permissible emission levels" (Russian abbreviation PDV) are determined by calculating the total

Table 3. Environmental Quality standards for Air (μgm^3)

Pollutant	EEC Period	EEC Concentration	USA Period	USA Concentration	USSR Period	USSR Concentration
Smoke	Yearly median of daily values	80			Daily arithmetic mean	50
	Winter median of	130				
SO_2	Yearly median of daily values	80-120	Annual arithmetic mean	80	Daily arithmetic mean	50
Particulates			Annual geometric mean	75	Daily arithmetic mean	150
CO			8-hour limit	10,000	Daily arithmetic	300
Ozone			1-hour limit	240	Daily arithmetic	30
NO_x			Annual arithmetic	80	Daily arithmetic	40(NO_2), 60(NO)

contribution of all polluting sources to a regional PDK for a given environmental parameter, and then defining a norm for the individual point source in such a way that the sum total of PDV will not exceed the mandated PDK. The determination of required factory emission levels must be reached by consensus among the enterprises, environmental agencies, and government authorities within the individual town.

According to the Law, new and reconditioned enterprises that do not meet air quality requirements may be denied authorisation to begin operations, and facilities violating norms may be closed down. However, the majority of enterprises cannot meet the costs of modern abatement equipment and it appears to be even more difficult to install the modern control devices in old factories, many of which were built some fifty or more years ago. This applies especially to the traditionally industrial areas, such as the Urals, where a large number of current enterprises originate from the last century. As far as new enterprises are concerned, the main goal has always been to complete a project in the least possible time and at the least possible cost, the environmental impact of these rushed projects and the long term effects on the environment being seldom calculated. In these cases, although the Law has set quite tough conditions, later decisions have compromised these, giving up environmental concerns to fulfil production quotas. Thus, the inventory of 1988 showed that at the Kazakh city of Ekibastuz which was known to produce the highest volume of dust emissions in the former Soviet Union (490 thousand tonnes in 1988, equal to the one eleventh of the USA total dust emissions from stationary sources for the same year), 90% of the pollution control equipment installed at the power plants burning coal was out of order (Bulletin, 1989). However, no measures leading either to immediate repair of the equipment or shutdown of the plants followed.

Another example is the construction of a copper-nickel smelting complex in Norilsk (Eastern Siberia, 69oN, 85oE). The city of Norilsk and the Norilsk Mining and Metallurgical Combine were established as part of Second five-year Plan (1932-36), which emphasised the domestic production of non-ferrous metals and ferro-alloys, and were expanded in the early 1970s. In 1990, emissions from Norilsk smelter totalled 2300 thousand tonnes, 2200 thousand tonnes of which was sulphur dioxide (about two thirds of total UK industrial SO_2 emissions). That makes Norilsk the worst Soviet city in terms of air pollution emissions. The annual ambient concentration of SO_2 was 140 mgm^{-3}, daily means reached 1.34 mgm^{-3} (Bulletin, 1991). Not only is air quality within the city extremely poor, but ecosystems show damage at a considerable distance from the smelter. The damage to forests is growing: in 1970 the area covered by completely dead forests was 6 thousand hectares; in 1988 it was 382 thousand hectares. Despite the situation being so desperate, no measures have been undertaken against the smelter or the Ministry for Non-ferrous Metallurgy which is responsible for the smelter.

Since many regions and urban areas would not be able to meet PDV (maximum permissible emission levels) and hence PDK, it was proposed by the industrial ministries that a progressive schedule be worked out whereby the industries in a given area would gradually bring their point-source emissions to the standard level. The interim standard was called the "temporarily agreed-upon emission" (Russian abbreviation VSV). Both PDV and VSV are the result of discussion and compromise, and at present they are based more on what industry says can be done than on any objective criteria. Unlike in the USA, in the Soviet Union strict time limits during which the necessary equipment must be installed, are either not set or may be reviewed. This constant compromising has made it impossible to enforce the approved principles of environmental legislation, although they have been transferred into concrete programmes, procedures and the highest environmental goals. Moreover, it has often resulted not only in high ambient levels of pollution but in the serious accidents. For example, at the Kirovograd copper smelter, where production of the chloridesulphone acid was started in 1941 at the improper building, no action was taken following the warnings made by the technical and environmental inspectorates. The result was a serious accident in 1989, when the massive emission of a poisonous gas completely destroyed more than 500 gardens belonging to local people, and, more seriously, affected two people (Proekt, 1991). Another example is the accident which took place at a Dagestan oil well in April 1990. In this case, environmental agencies did not manage to prevent the violations in the abatement equipment usage that allowed the massive emission of the gas condensate which lasted for six days, about 200-300 thousand m3 being emitted per day (*ibid.*).

Enforcement of the legislation

Environmental offences were included in the criminal codes of the Republics in the 1960s and have not been reviewed or reformulated since then, resulting in numerous uncertainties.

During the 1980s, more than 3,000 environmental offences were registered each year, the biggest share dealing with illegal hunting and a much smaller number with air and water pollution. Both criminal and administrative sanctions can follow a violation of environmental regulation, however, the legal process has never been central to conflict resolution, administrative processes, both formal and informal, being more important than litigation. Administrative sanctions range that from moral sanctions to closing down the offending enterprise, loss of employment, disciplinary action, and restitution of damage. The most common practices, however, were fines and penalties. In 1976, personal liability was set at 10.000 roubles and enterprise liability as 100.000 roubles (Jancar, 1988). While 10.000 roubles might serve as a deterrent to individuals, the penalty for enterprises was very modest, particularly considering that the fines were normally included within the operating costs.

The emphasis on administrative measures provided the biggest difference between enforcement of environmental legislation in the West and in the former USSR. Central to Western law is the practice of litigation to force compliance with laws, to impose penalties for legal violations, and to settle disputes between parties having conflicts of interest. This makes polluters aware that legal actions can be initiated if they refuse to follow the directives. In the former Soviet Union, violators might frequently ignore environmental rules and be sure that their case would not make it to court. Thus in 1989-90, environmental violations made only 5% of the cases submitted to arbitration (Proekt, 1991).

The question of damage liability also contributes to the weakness of enforcement provisions in the environmental area. The law is not clear in its definition of what constitutes "damage", what are the economic costs incurred, or who is the legal property owner or user.

Environmental agencies

Another cause of problems lies in the diversity of agencies involved in environmental protection, and the lack of co-ordination in their efforts.

Thus before 1988, 15 agencies were authorised to monitor air pollution (Juncar, 1988). The US Information Agency Soviet Research Institutes Project identified a total of 329 research establishments in the USSR dealing with some aspects of the environment (ibid.). There is no doubt that a large country such as the USSR should have a sizeable number of agencies involved in matters of environmental protection, but as it has been frequently suggested by Soviet specialists, there were too many of them, too much fragmentation and too little co-ordination in the Soviet environmental management efforts.

Often the enforcement functions lay within the various ministries which had been largely the cause of the problem in the first place. For example, the former State Agro-Industrial Committee was responsible for the increase of agricultural production on the one hand and for the control of fertiliser runoff on the other; the former Ministry of Water Resources and Land Reclamation - for development of irrigation projects and promotion of water conservation. In both cases the former issues were held to be more significant that the latter ones. The example from the area of air pollution is the Ministry of Non-ferrous Metallurgy which was responsible both for production and for ensuring compliance with standards.

The position of "environmental" ministries and industrial enterprises in the Soviet hierarchy was different. Two types of ministries - all-union (submitted to the central government only) and republican (submitted to the republican governments) - existed, all-union ministries being much more important. For example, such industries as iron and steel, chemical, or non-ferrous metallurgy which have a very high potential for air pollution, and which are widely distributed throughout the country, were accountable only to the all-union ministries and to the central government. "Environmental" ministries (those ones which included departments responsible any kind of environmental preservation), being mainly republican ones, were of a lower status had therefore no power against the industrial enterprises. This enabled industrialists to ignore environmental directives safely. Even if the polluting ministries and a regulating body (as, for example, the Sanitary Epidemiological Service attached to the all-union Ministry of Health) had an equal status, the polluting ministries had the option of negotiation with the environmental agencies only at the top of the hierarchy at the level of the USSR Council of Ministries instead of at the local level.

This separation of the central and republican (local) responsibilities was the other important reason impeding the implementation of the environmental laws.

A solution to the dispersal of administrative and exploitative functions has been to delegate the co-ordination of environmental management to a lead agency. American experience of creating the rather powerful Environmental Protection Agency was a good example. In 1978, an attempt was made to centralise the responsibility for monitoring state of the Soviet environment: the national Hydrometeorological Service was elevated, expanded, and restructured as the State Committee for Hydrometeorology and the Environment. Its new tasks included the monitoring of air quality throughout the country, inspecting factories and construction sites, and assessing the degree of compliance with the corresponding environmental regulations and standards (Kolbasov, 1983). Despite the limitation of sharing jurisdiction of environmental matters with other agencies and branch institutions, the elevation brought the new State Committee an organisational permanence and some authority. Unlike all other Soviet institutions involved in environmental protection, the State Committee on Hydrometeorology and the Environment did not have resource exploitation as its primary function; it did not necessitate a choice between the assertion of environmental and resource use priorities.

Although the Committee's charge was broad, its focus soon became concentrated, mainly on monitoring air pollution. In this case, a certain success has soon become evident. The national network of air quality monitoring, set up in 1972, increased significantly (table 4). In some cities (Moscow, St. Petersburg, Kiev, Rostov, Ekaterinburg, etc.) the pollution monitoring system is particularly advanced. Normally, four main pollutants

Table 4. Number of Cities Submitting Information about Emissions And Air Quality to the State Committee on Hydrometeorology and the Environment

Year	1980	1983	1984	1985	1986	1987	1988	1989	1990
Emission values	534	422	502	544	567	599	600	-	-
Ambient air quality	442	450	457	458	476	514	537	560	579

(suspended particles, sulphur dioxide, nitrogen dioxide and carbon monoxide) are being monitored. However, in some areas subject to the specific kinds of pollution, others may be under observation. Temporary sites may be installed near the most dangerous sources. Under the weather conditions most favourable for accumulation of pollutants, special actions may be undertaken. Thus in June 1984, in Tashkent, the monitoring of pollution concentrations resulted in an "alert" being proclaimed and in prompt action being undertaken to reduce emissions (Elsom, 1985). Industrial plants were instructed to reduce emissions to prearranged lower levels, asphalt boilers on construction sites were shut down for several hours, and road works raising dust were allowed only at night until the pollution episode had abated. In Moscow, the Hydrometeorological Centre daily forecasts the meteorological conditions influencing air quality. The recommendations are passed to the city council which then decides whether emissions should be decreased.

However, organisation of the monitoring network suffers from certain drawbacks. Firstly, the rather advanced systems are normally located in the big administrative or industrial centres. However, heavily polluting enterprises are very often located in rather small industrial towns. The best example is the city of Asbest in the Urals, where one of the biggest asbestos producing plants is located. The level of extremely dangerous asbestos containing dust emissions has been growing steadily since 1980 and reached some 250 thousand tonnes a year in 1988. However, monitoring of ambient air quality in the city was started only in 1990. In other small Urals city Troitsk where the emissions of suspended particles alone exceeded 320 thousand tons in 1988, air quality monitoring has not been started yet. Therefore, the Committee has had only limitted success restricted mainly by monitoring of air pollution.

In 1988, it was decided to relegate the Committee for Hydrometeorology back to scientific research organisation and monitoring. To carry out its environmental management duties, a new agency, the USSR State Committee for Environmental Protection and Rational Use of Natural Resources, was formed. It was installed as a union-republic ministry, which meant that each of the USSR's 15 Republics would have a similar local organisation. The new agency's tasks, which were delegated from a number of "environmental" ministries and committees, were described as carrying out the comprehensive environmental protection activities of the country: environmental monitoring; long-range environmental planning; promulgating norms and standards; overseeing the design, siting, and construction of environmentally sensitive facilities; issuing waste disposal permits; managing nature preserves, endangered species, and hunting; environmental education and industrial co-operation and co-ordination (Pryde, 1990). This is a huge agenda, and would have required a powerful agency to implement it. Unfortunately, the Committee has not been given the necessary funding, authority and personnel to successfully carry out all those vitally important charges; despite its high position in the ministerial hierarchy its importance, as well as the importance of its forerunner was low.

Environmental issues after 1988

The most dramatic change in Soviet environmental affairs in recent years was the expansion of information available to the public about the actual condition of the environment. Whereas previously data about pollution levels and health effects of pollution were available only to scientists and governmental officials, now they are broadly published in the mass media.

The condition of the natural environment has become an increasingly important political issue in the USSR. Thus, in 1989-90, public protests against plant siting, ecological platforms as components of minority nationality programs, and public discussion of adverse health effects became everyday occurrences. Environmental issues formed an important component of election platforms of numerous candidates for the Congresses of Peoples' Deputies. As a result of this political activism, many major projects have been suspended or cancelled. Thus, operation of the Astrakhan' Gas Condensate Complex was suspended; some nuclear plants were shut down and construction of others was stopped (DeBardeleben, 1990). In 1988, about 130,000 people addressed the State Committee for the Environmental Protection demanding that it shut down a number of power plants, chemical plants, etc. One of the letters, from the town of Salavat (Baskiria) which is a centre for chemical industry and oil refining was signed by 50,000 people; the protest against constructing of a power plant in Rostov was signed by 20,000 people (Proekt, 1991).

There are some positive signs in the development of air quality regulations. The Russian Ministry for Environmental Protection (the former State Committee for Environmental Protection) is determined to change the way it charges violating enterprises. Thus, from 1993 all new enterprises located within the Russian Federation will be obliged to pay for the produced pollution from its profits. If a violating enterprise is not able to cover its costs, the Ministry will be able to close it. This strategy seems more realistic now, as many of the newly opened enterprises are private and there will be no "ministerial pressure" on the environmental agencies. As far as already operating enterprises are concerned, charges for the emissions which cannot be reduced without the installation of the equipment the enterprise cannot afford, will be included into the operation costs; charges for the emissions above this will be paid from the profit. As the main purpose is not to collect the fines, but to improve the air or water quality, the enterprises will be able to claim the money back if they intend to spend it on pollution control equipment ("Delovoi mir", 1992). The first attempt to introduce these regulation were made already in 1990 in the Baikal region (Potravny et al., 1992), and resulted in a considerable increase of the investments to the pollution control equipment. However, despite some positive signs economic priorities have changed only slowly. Numerous plants closures or cancellations of construction projects based on ecological concerns (closing down of an Armenian nuclear station; suspension of the third block construction at the Ignalina station in Lithuania; conservation of the plant in Chapaevsk, slated to destroy chemical weapons; closing of the Novo-Kokand fertiliser plant in Usbekistan; etc.) are associated with the tremendous costs. Moreover, research experience in other countries suggests that it is necessary to spend approximately 2% of GNP for the stabilisation of ecological conditions (Robins, 1991). The well-known Russian environmentalist A. Yablokov stated that "in order to improve the quality of the environment what must be spent in the USSR is no less than 5%. We are still spending around one" (Lavrov, 1990). The scarce financial resources available for attaining the environmental goals of the USSR are a factor that will be in effect for a rather long time. Thus, environmental problems were not even mentioned in one of the most popular programmes of economic reforms "the 500 days" ("Delovoi mir", 1992). More urgent problems such as housing, the health service, and food supply play the main role; when the traditional argument "employment-versus-environment" is used, the decision is still normally not in favour of the environment. It is no surprise that there is a certain evidence that environmental activity has been decreasing in 1990-1992 after the peak of 1988-1989. In his interview to the newspaper "Delovoi mir" the Minister of Environment, Dr.Danilov-Danilian, characterised the state of environmental movement in 1991-1992 as lethargy.

The collapse of the Soviet Union has aggravated these environmental problems, many of which have now become international. Numerous political and ethnic conflicts, and further decreases of production suggest that, in the face of other serious economic problems, the capability to address environmental problems effectively may not be improving. The number of accidents (table 5) which has been increasing dramatically in the majority of Republics already in 1989-90 support this sad conclusion.

Table 5. Accidents Damaging Environment in 1989 and 1990

Republic	**Number of Accidents**		**Damge (million rubles)**	
	1989	1990	1989	1990
USSR	1754	3525	231.8	338.3
Azerbaidzan	16	50	8.8	6.0
Armenia	80	19	4.4	0.4
Belorussia	14	29	1.0	4.8
Georgia	19	8	1.2	2.0
Kazakhstan	41	92	6.3	7.5
Kirgizia	11	6	0.1	0.2
Latvia	9	141	0.2	1.6
Lithuania	no data	no data	no data	no data
Moldavia	73	45	2.6	1.2
Russia	1173	2659	185.4	285.9
Tadzikistan	2	3	0.05	0.04
Turkmenia	4	6	0.4	0.2
Usbekistan	34	50	1.1	3.7
Ukraine	293	393	20.0	22.2
Estonia	22	24	0.1	2.6

Conclusions

The system of environmental (including air quality) management in the former Soviet union was notorious for its poor enforceability. There were clear reasons which allowed polluters to feel safe despite non-compliance. The main one was the centrally planned economy which forced enterprises to produce at any cost. In this permanent conflict between production and protection, the former always had a priority. The system of non-compliance fees excused enterprises from making a sufficient abatement effort. The diverse system of environmental agencies and their low priority frustrated their efforts. Theoretically, these factors are fading away now; however, practically the environmental concerns have not been given sufficient economical priority and the highly corrupted system makes enforceability of the new environmental regulations questionable. There are a number of positive signs such as a reduction in emissions and some improvement of air quality; however, the main reason for this is the decrease of production rather than environmental measures. Further decline of industrial output, the general collapse of the economy, and numerous ethnic and military conflicts all provide grounds for the conclusion that the environmental recovery is still far away.

For membership forms, or further information about IRNES,please contact:

Rachel Reeve
School of Geography
University of Oxford
Mansfield Road
Oxford
OX1 3TB